THE ABSOLUTE RELATIONS OF TIME AND SPACE

THE ABSOLUTE RELATIONS OF TIME AND SPACE

BY

ALFRED A. ROBB, Sc.D., D.Sc., Ph.D.

CAMBRIDGE
AT THE UNIVERSITY PRESS
1921

CAMBRIDGE
UNIVERSITY PRESS

University Printing House, Cambridge CB2 8BS, United Kingdom

Cambridge University Press is part of the University of Cambridge.

It furthers the University's mission by disseminating knowledge in the pursuit of education, learning and research at the highest international levels of excellence.

www.cambridge.org
Information on this title: www.cambridge.org/9781107536807

First published 1921
First paperback edition 2015

A catalogue record for this publication is available from the British Library

ISBN 978-1-107-53680-7 Paperback

PREFACE

At the meeting of the British Association in 1902, Lord Rayleigh gave a paper entitled "Does motion through the ether cause double refraction?" in which he described some experiments which seemed to indicate that the answer was in the negative. I recollect that on this occasion Professor Larmor was asked whether he would expect any such effect and he replied that he did not expect any.

In the discussion which followed reference was made to the null results of all attempts to detect uniform motion through the aether and to the way in which things seemed to conspire together to give these null results.

The impression made on me by this discussion was: that in order properly to understand what happened, it would be necessary to be quite clear as to what we mean by equality of lengths, etc., and I decided that I should try at some future time to carry out an analysis of this subject.

I am not certain that I had not some idea of doing this even before the British Association meeting, but in any case, the inspiration came from Sir Joseph Larmor, either at this meeting or on some previous occasion while attending his lectures.

Some years later I proceeded to try to carry out this idea, and while engaged in endeavouring to solve the problem, I heard for the first time of Einstein's work.

From the first I felt that Einstein's standpoint and method of treatment were unsatisfactory, though his mathematical transformations might be sound enough, and I decided to proceed in my own way in search of a suitable basis for a theory.

In particular I felt strongly repelled by the idea that events could be simultaneous to one person and not simultaneous to another; which was one of Einstein's chief contentions.

This seemed to destroy all sense of the reality of the external world and to leave the physical universe no better than a dream, or rather, a nightmare.

If two physicists A and B agree to discuss a physical experiment, their agreement implies that they admit, in some sense, a common world in which the experiment is supposed to take place.

It might be urged perhaps that we have merely got a correspondence between the physical worlds of A and B, but if so, where, or how, does this correspondence subsist?

It cannot be in A's mind alone, or it would not be a correspondence, and similarly it cannot be in B's mind alone.

It seems to follow that it must be in some common sub-stratum; and this brings us at once back to an objective standpoint.

The first work which I published on this subject was a short tract entitled *Optical Geometry of Motion, a New view of the Theory of Relativity* which appeared in 1911.

This paper, though it did not claim to give a complete logical analysis of the subject, yet contained some of the germs of my later work and, in particular, it avoided any attempt to identify instants at different places. Later on the idea of "*Conical Order*" occurred to me, in which such instants are treated as definitely distinct.

The working out of this idea was a somewhat lengthy task and in 1913 I published a short preliminary account of it under the title *A Theory of Time and Space*, which was also the title of a book on this subject on which I was then engaged.

This book was in the press at the time of the outbreak of the war and was finally published towards the end of 1914.

Unhappily at that period people were concerning themselves rather with trying to sever one another's connexions with Time and Space altogether, than with any attempt to understand such things; so that it was hardly an ideal occasion to bring out a book on the subject.

The subject moreover was not an easy one, and I have been told more than once that my book is difficult reading.

To this I can only reply as did Mr Oliver Heaviside, under similar circumstances, that it was perhaps even more difficult to write.

Be that as it may, the results arrived at fully justified my attitude towards Einstein's standpoint.

I succeeded in developing a theory of Time and Space in terms of the relations of *before* and *after*, but in which these relations are regarded as absolute and not dependent on the particular observer.

In fact it is not a "theory of relativity" at all in Einstein's sense, although it certainly does involve relations.

These relations of *before* and *after*, serving, as they do, as a physical basis for the mathematical theory, were quite ignored in Einstein's treatment; with the result that the absolute features were lost sight of.

Even now, some six years from the date of publication of my book, comparatively few of Einstein's followers appear to realize the extreme importance of these relations, or to recognize how they alter the entire aspect of the subject.

The theory, in so far as its postulates have an interpretation, becomes a physical theory in the ordinary sense, but these postulates are used to build up a pure mathematical structure.

From the physical standpoint the question is: whether the postulates *as interpreted* are correct expressions of physical facts, or in some respect only approximations?

If the postulates are not all correct expressions of the facts, then which of them require emendation and what emendation do they require?

As regards the pure mathematical aspect of the theory: this of course remains unaffected by the physical interpretation of the postulates, and those who are interested only in pure mathematics may find that the method employed has certain advantages as a study of the foundations of geometry.

In particular it may be noticed that by this method we get a system of geometry in which "congruence" appears, not as something extraneous grafted on to an otherwise complete system, but as an intrinsic part of the system itself.

I had intended making further developments of this theory, but the outbreak of the war caused an interruption of my work.

In the meantime Einstein produced his "generalized relativity" theory and the reader will doubtless wish to know how this work bears upon it.

So far as I can at present judge, the situation is this: once coordinates have been introduced, the theory here developed gives rise to the same analysis as Einstein's so-called "restricted relativity" and this latter cannot be regarded as satisfactory apart from my work, or some equivalent.

Einstein's more recent work is extremely analytical in character. The *before* and *after* relations have not been employed at all in its foundation, although it is evident that, if these relations are a sufficient basis for the simple theory, they must play an equally important part in any generalization. Moreover these relations most certainly have a physical significance whatever theory be the correct one.

A generalization of my own work is evidently possible and, to a certain extent, I can see a method of carrying this out, although I have not as yet worked out the details. (See Appendix.)

In the meantime it seemed desirable to write some sort of introduction to my *Theory of Time and Space* which, while not going into the proofs of theorems, would yet convey to a larger circle of readers the main results arrived at in that work.

A. A. R.

CAMBRIDGE,
November 12, 1920.

CONTENTS

THE ABSOLUTE RELATIONS OF TIME AND SPACE

PRELIMINARY CONSIDERATIONS

THE study of Time and Space is one which in certain respects is extremely elusive and involves a number of difficulties which in ordinary daily life we are apt to overlook.

In scientific work, however, it is all-important to have clear ideas and to know exactly what our statements mean.

This is by no means always an easy task, for it frequently happens that our crude ideas on certain things may be sufficiently precise for certain purposes, but not precise enough for others.

Thus in the ordinary elementary teaching of plane geometry there are certain difficulties which are generally passed over, largely because they are real difficulties and a proper understanding of them could hardly be expected from a beginner.

For instance the use of ruler and compasses and the method of superposition.

The use of the ruler conveys a somewhat crude idea of what we mean in the physical world by points lying in a straight line, while the use of compasses conveys an equally crude idea of what we mean by points in a plane being equally distant from a given point in the plane.

The method of superposition involves ideas which are closely akin to those involved in the supposed use of compasses, but of a more elaborate character.

Both sets of ideas may be described as *ideas of congruence.*

Although there are other difficulties besides these to be overcome, still these will suffice for our present purpose, which is to show that certain points have been slurred over when we first began the study of geometry, which later on may require further elucidation.

Now let us approach this subject as a beginner of sufficient intelligence might be supposed to do.

There is one thing which we might observe, namely: that though we make use of figures drawn on paper to assist us in keeping the facts in mind, yet in proving a theorem, as distinguished from making use of the result, there is no necessity that the figure should be accurately drawn. A very rough figure will suffice and, if we are fairly expert, and the theorem not too complicated, we can dispense with a figure altogether.

Next let us suppose the figures to be accurately drawn on a plane sheet of paper (whatever the expressions "accurately drawn" and "plane" may mean) and then suppose the sheet of paper to be rolled up into a spiral, we could still make use of the figures on the curved sheet as mental images in proving our theorems, although our original straight lines would now (with certain exceptions) be no longer straight.

We could however substitute for our ruler a flexible string, drawn taut, so as to lie in contact with the curved surface of the paper and similarly we could make use of a flexible inextensible tape line or string instead of our original compasses and all our theorems would work out as before, except that lines would be curved which had originally been straight and lengths would be measured along such lines instead of "directly" between points.

With such modifications, to every theorem concerning figures on the plane sheet there will be a corresponding theorem concerning figures on the curved sheet and *vice versa*, and similar methods of proof may be employed in the two cases.

Though the objects about which we are reasoning in the two cases are different, yet the logical processes are formally the same.

We can, however, go still further and consider the case where the figures are accurately drawn (whatever that may mean) on a plane sheet of india-rubber which is then stretched in any way.

In this case straight lines on the unstretched rubber would become lines, straight or curved, on the stretched rubber and a closed curve such as a circle would remain closed after the stretching.

Further, curves which intersected would still intersect and curves which did not intersect would not intersect after the stretching.

A point which lay inside a closed curve such as a circle, would become a point inside a closed curve on the stretched rubber.

Again, a point which lay between two other points in a line of some sort on the unstretched rubber would become a point between two corresponding points on the corresponding line on the stretched rubber.

The distances between the points would of course have altered according to our original standard and two lengths which were originally equal might no longer be equal, but nevertheless certain correspondences would still hold and could be traced between theorems involving equality of lengths on the unstretched rubber and theorems on the stretched rubber.

Perhaps the simplest way of seeing this is to introduce a system of coordinates (say Cartesian coordinates) on the unstretched rubber, by which any point of it would be represented by two numbers.

If then we imagine the rubber to be stretched, the same pairs of numbers could be taken to represent the same points of the rubber after stretching as before. The axes would now, generally speaking, become curved lines and the parallels to them would also in general become curved lines.

The points equidistant from a given point on the unstretched rubber would lie in a circle, and if the equation of this circle be taken as

$$(x - a)^2 + (y - b)^2 = r^2,$$

then this equation would represent also some curve on the stretched rubber. The radii of the circles would become some sort of lines all passing through one point and intersecting the distorted circle.

We should in this way get lines which had been straight, curves which had been circles, lengths which had been equal, etc., and we could deal with these algebraically in the same way as we did with the straight lines, circles and equal lengths on the unstretched rubber.

We notice that the things which actually do remain permanent are the particles of the rubber and certain features of their order.

If we consider the coordinate system we observe that, although the axes and the parallels to them are in general no longer straight after the stretching, yet as either set of parallel lines did not intersect before stretching, so the corresponding lines do not intersect after stretching and they preserve their original order.

We know however that, after a proper foundation has been laid, any geometrical theorem may be proved by coordinate methods and so it is evident that all reasoning which is done after coordinates have once been introduced will apply equally in dealing with certain other things than lines which are truly straight and lengths which are truly equal.

Thus though the sheet of rubber may have originally been plane, yet after stretching it may be curved in innumerable different ways and yet there are certain features which remain invariant throughout.

It is thus evident that although for purposes of mathematical reasoning the actual straightness of lines or actual equality of lengths in the ordinary sense of the terms is not essential, yet when we wish to make use of geometry to describe the physical world the meanings of "straightness" and "equality of length" are all-important.

It is not sufficient that we should say that "there are such things as straight lines," or that "there are lengths which are equal," but it is necessary to have criteria by which we can say (at least approximately) "here are points which lie in a straight line" and "here is a length which is equal to yonder length."

The ruler and compasses give us rough standards of straightness and equality of length in the sense in which these terms are used in ordinary life, but, if we wish to go in for extreme accuracy, other standards must be employed and we must get more precise ideas as to what we really wish to convey when we make use of such expressions.

Consider first the question of what we mean when we say that two bodies are of equal length.

The ordinary method of comparing them is to make use of a measuring rod which we regard as *rigid*; or an *inextensible* tape line. But what do we mean by these words "*rigid*" and "*inextensible*"?

We find that it is by no means easy to say exactly what we do mean.

Approximate rigidity and inextensibility are common enough properties of solid bodies, but by the application of force all bodies are found to be more or less elastic, while change of temperature will also change the length of a rod compared with a parallel rod.

Again, if we wish to compare lengths which are not parallel, the usual mode of procedure would be to turn a measuring rod round from parallelism with the one length into parallelism with the other.

The possibility then arises that during the motion the standard may alter and give us results which indicate the lengths as equal when in reality (whatever that may mean) they are different.

Thus for example, if we wished to compare the lengths OA and OB where A and B are, say, the extremities of the major and minor axes of an ellipse whose centre is O, and suppose we had an elastic tape line which we place first with one end at O and the other at B.

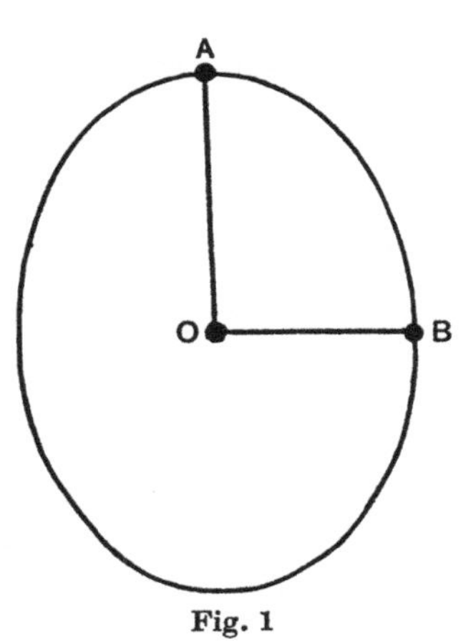

Fig. 1

If then keeping the one end fixed at O we move the other round the ellipse we should apparently get the same length for OA as for OB.

Now although this seems fantastic, yet the famous experiment of Michelson and Morley seemed to show that just this sort of thing did happen when a body was turned round from a position such that its length was parallel to the direction of the earth's motion in its orbit into a position such that its length was perpendicular to that direction.

The experiment, which was an optical one, consisted in dividing a beam of light into two portions which travelled, the one in one direction, and the other in a transverse direction, and were reflected back again by mirrors.

If we adopt ordinary ideas for the moment and suppose the light to be propagated with a velocity v through a medium and the apparatus to move through that medium with a velocity u, it is easy to calculate the time of the double journey for the two portions of the beam.

For the case of a part of the beam which travels in the direction of motion of the apparatus and back again the time of the double journey is found to be

$$t_1 = \frac{2va_1}{v^2 - u^2},$$

where a_1 is the distance between the point of the apparatus where the beam divides and the corresponding reflector. For the case of the transverse portion of the beam the time of the double journey is found to be

$$t_2 = \frac{2a_2}{\sqrt{v^2 - u^2}},$$

where a_2 is the distance between the point of the apparatus where the beam divides and the other reflector.

Now it is possible to arrange things so that $t_1 = t_2$ and this can be done with extreme accuracy by means of the interference bands which are produced.

We should then have

$$\frac{2va_1}{v^2 - u^2} = \frac{2a_2}{\sqrt{v^2 - u^2}},$$

giving

$$a_1 = \sqrt{1 - \left(\frac{u}{v}\right)^2}\, a_2.$$

Thus a_1 would be slightly less than a_2.

It was found however that, when the apparatus was caused to rotate at a uniform slow rate, and the times of the double journey were equal for one position of the apparatus, then they were equal for all positions. This seemed to indicate that the dimensions of the apparatus in different directions changed as it rotated and the view was put forward by Fitzgerald and Lorentz that a material solid body contracted in the direction of its motion so that a sphere moving through space with a velocity u became a spheroid whose major and minor axes were in the ratio

$$1 : \sqrt{1 - \left(\frac{u}{v}\right)^2},$$

where v is the velocity of light.

It is clear that this once more raises the question as to the real meaning of "equality of length" from which we started out.

Solid bodies apparently do not provide us with standards sufficiently permanent for dealing with such problems.

But the subject of motion raises a number of other difficulties.

There is in particular the question of "absolute motion" and whether this expression has any precise meaning.

The underlying idea of those who believe in "absolute motion" is that, if we consider a definite point of space at any instant, then that point preserves its identity at all other instants. The difficulty of identifying a point of space at two different instants is freely admitted, but for all that (so it is contended) the identity persists.

It was however noticed that, in the classical Newtonian Mechanics, the equations of motion preserved the same form for a system of bodies whose centre of inertia was in uniform motion in a straight line as for a similar system whose centre of inertia was "at rest," so that purely mechanical phenomena could not be expected to show up any difference between the two cases.

The question then naturally arose whether any difference could be detected by optical or electrical means, but experiment failed to show any.

Nextly it was pointed out by Larmor and Lorentz that the electromagnetic equations could also be transformed by a linear substitution so that they preserved the same form for a system moving with uniform velocity as they had for a system "at rest."

In order to do this, however, a "*local time*" had to be introduced.

We are all familiar with the use of "local time" on the earth's surface, but the cases are different in one important respect.

The idea underlying the use of "local time" on the earth's surface is simply that of having different names in different parts of the world for what is regarded as the same instant. Thus noon at Greenwich and noon at New York are both described as 12 o'clock local time, although the instants referred to are clearly different. On the other hand the use of chronometers in navigation is regarded as a method of approximately identifying the same instant at different parts of the earth.

But, as previously remarked, the "local time" used in transforming the electromagnetic equations is of a different character and events which are regarded as simultaneous according to one "local time" would not be simultaneous, in general, when compared by the "local time" of a system which was in motion with respect to the first.

We might of course regard the one "local time" as the true time and the other as a mathematical fiction, but there is no reason known why we should select the one rather than the other, just as there is no way of distinguishing a body "at rest" from one moving uniformly in a straight line.

In fact it appears that, just as we have no method of distinguishing the same point of space at two distinct instants of time, so we cannot strictly identify the same instant of time at two distinct points of space.

It is to be observed that though we started out by trying to give a precise meaning to the idea of equality of length, in which we seemed to be concerned only with space, yet in our attempt to do so, we find difficulties with regard to time intruding themselves.

We can see, however, that even in our original use of compasses the time element intrudes, since in comparing lengths by the use of compasses, the compasses are moved and the idea of motion involves that of time.

Also in the Michelson and Morley experiment, since light takes a finite interval of time in getting from an object P to an object Q and back again to P, we are introducing time relations in comparing lengths.

The question now arises: suppose we imagine a flash of light sent out at an instant A from a particle P to a distant particle Q and arriving there at an instant B and suppose it reflected back to P where it arrives at an instant C; how are we to identify the instant B with any instant at P between A and C?

If we regard P as being "at rest" we might reasonably think to identify B with the instant at P which is midway between A and C, but this would imply that we had some means of measuring intervals of time, and that brings us up against all the same sort of difficulties which we encountered in trying to find a satisfactory method of measuring space intervals.

On the other hand, if P be in uniform motion in the direction PQ it would seem that B is not identical with the instant at P which is midway between A and C.

In any case we do not know of any means of telling whether P is "at rest" or not.

Having thus been led on from the consideration of spacial relations to those of time we seem at first sight to have increased our difficulties instead of solving them, but if we persevere in our task we shall find that we have made an appreciable advance towards solving our problem.

From the consideration of figures drawn on a sheet of rubber which was afterwards stretched in any way, we were led to recognise the importance of *order* in the study of the logic of geometry, and since order also plays a part in time relations, it seems worth while to consider order in time.

Now here we find an interesting and very important thing.

If I consider two distinct instants of which I am *directly conscious**, I notice that the one is *after* the other.

Noon to-day is *after* noon yesterday and I cannot invert the order.

There is in fact what is called an asymmetrical relation between the two instants, such that if B be *after* A, then A is not *after* B.

* The fact that I am *directly conscious* of the two instants is very important, in view of later developments.

If we consider two points or two particles in space, say P and Q, there is nothing analogous to this and we have no reason to say that Q is *after* P rather than that P is *after* Q.

We might, of course, give them an order by means of some convention, but such convention would be quite arbitrary, whereas in the case of the instants, it is a matter of fact and not of convention, quite independently of what words we may employ to express that fact.

The simplest relation of order among points is a relation of *between* which involves three terms instead of two.

This relation of *between* has been employed by various mathematicians in investigating the foundations of geometry, but the relation of equality of lengths then appears as something extraneous, grafted on to the system.

The use of an asymmetrical relation such as *after* appears to have great advantages over a relation such as *between* in constructing a theory of order and I have found it possible, by means of such a relation, to construct a system of geometry of space and time. It might perhaps more correctly be described as a geometry of time, of which spacial geometry forms a part.

In constructing this system it is necessary to modify certain currently accepted notions, but the modifications required all appear to be capable of justification and the structure, when completed, will be found closely to resemble our ordinary conceptions.

We shall regard an instant as a fundamental concept which, for present purposes, it is unnecessary further to analyse, and shall consider the relations of order among the instants of which I am directly conscious.

Thus for such instants we find the following properties:

(1) If an instant B be *after* an instant A, then the instant A is not *after* the instant B, and is said to be *before* it.

(2) If A be any instant, I can conceive of an instant which is *after* A and also of one which is *before* A.

(3) If an instant B be *after* an instant A, I can conceive of an instant which is both *after* A and *before* B.

(4) If an instant B be *after* an instant A and an instant C be *after* the instant B, the instant C is *after* the instant A.

(5) If an instant A be neither *before* nor *after* an instant B, the instants A and B are identical.

The set of instants of which I am directly conscious have thus got a linear order.

But now let us consider the fifth of these properties.

It might at first sight be supposed that it was a necessary consequence of our ideas of *before* and *after*. That it is really logically independent of the other properties may be shown by the help of a geometrical illustration. This illustration is very suggestive and we purpose to make further use of it, but the logic of our theory is independent of the illustration.

Suppose we have a set of cones having their axes parallel and having equal vertical angles, and further, suppose each cone to terminate at the vertex, which is however to be regarded as a point of the cone.

We shall call such a cone having its opening pointed upwards an α cone, and one with the opening pointed downwards a β cone.

Thus corresponding to any point of space there is an α cone of the set having the point as vertex, and similarly there is a β cone of the set having the point as vertex.

Now it is possible by using such cones and making a convention with respect to the use of the words *before* and *after* to set up a type of order of the points of space.

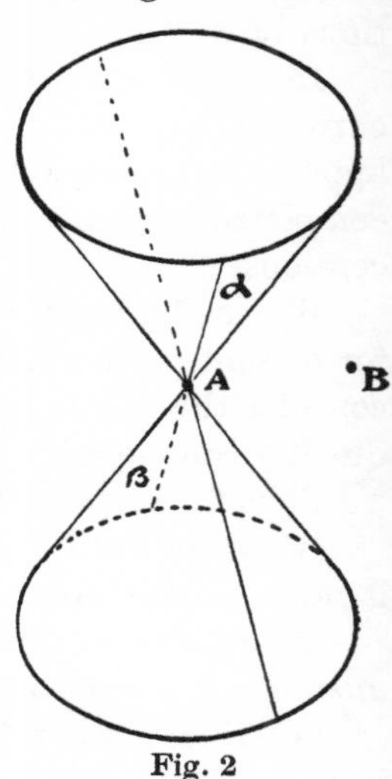

Fig. 2

For the purposes of this illustration we shall make the convention that, if A_1 be any point and α_1 and β_1 be the corresponding α and β cones, then any point A_2 will be said to be *after* A_1 provided it is distinct from A_1 and lies either on or inside the cone α_1 and will be said to be *before* A_1 provided it is distinct from A_1 and lies either on or inside the cone β_1.

It is easy to see that with this convention we have the following:

(1) If a point B be *after* a point A, then the point A is not *after* the point B and is said to be *before* it.

(2) If A be any point, there is a point which is *after* A and also a point which is *before* A.

(3) If a point B be *after* a point A there is a point which is both *after* A and *before* B.

(4) If a point B be *after* a point A and a point C be *after* the point B, the point C is *after* the point A.

We cannot however assert that if a point A be neither *before* nor *after* a point B, that the points A and B need be identical.

This is easily seen since the point B might lie in the region outside both the α and β cones of A. (Fig. 2.)

This illustration shows that the fifth condition is logically independent of the other four.

The type of order which we have illustrated by means of the cones, we shall speak of as *conical order*, but the logical development of the subject is independent of this illustration.

We may note however in passing that, if A and B be distinct points one of which is neither *before* nor *after* the other, then there are points which are *after* both A and B and also points which are *before* both A and B.

This follows since in this case the α cones of A and B intersect, as do also the β cones of A and B.

It should further be noted that if we have any line straight or curved in space, but whose tangent nowhere makes a greater angle with the axes of the cones than their semi-vertical angle, then if we confine our attention to the points of any one such line, we can assert that: if a point A be neither *before* nor *after* a point B, the points A and B are identical.

Thus provided we confine our attention to the points of such a line, the whole five conditions are satisfied.

Returning now to the consideration of instants, we observed that there was a difficulty in identifying the same instant at different places.

The relations of *before* and *after*, however, enable us to say in certain cases that instants at a distance are distinct. Thus if I can send out any influence or material particle from a particle P at the instant A so as to reach a distant particle Q at the instant B then this is sufficient to show that B is *after* and therefore distinct from A.

If now the influence or material particle be reflected back to P and arrives there at the instant C, then C is *after* and therefore distinct from B, while moreover, C is *after* A.

Now suppose the influence be a flash of light or other instantaneous electromagnetic disturbance and we appear to have reached a limit.

We do not seem to be able to send out any influence or material particle from P at any instant *after* A so as to arrive at Q at the instant B, and we do not seem to be able to send out any influence or material particle from Q at the instant B so as to arrive at P *before* the instant C.

In fact the range of instants at P which are *after* A and *before* C appear to be quite separated from the instant B so far as any influence is concerned.

Now let us suppose that light has this property.

It may or may not be strictly true of light but, provided there be some influence which has this property (and others which we shall specify later), such influence will serve for the purpose in hand, and we shall, provisionally at any rate, ascribe it to light.

Now B could at most be identical with only one of the instants at P, and such instant would require to be *after* A and *before* C, but we have no means of saying which instant it is.

The other instants in this range would then all be either *before* or *after* B.

But what now do we really mean when we say that one instant is *after* another or one event *after* another?

If I at the instant A can produce any effect however slight at the instant B, this is *sufficient* to imply that B is *after* A.

A present action of mine may produce some effect to-morrow, but nothing which I may do now can have any effect on what occurred yesterday.

It appears to me that we have here the essential features of what we really mean when we use the word *after*, and that the abstract power of a person at the instant A to produce an effect at a distinct instant B is not merely a *sufficient*, but also a *necessary* condition that B is *after* A.

If we accept this as the meaning of *after* it would then appear that no instant at P which is *after* A and *before* C is either *before* or *after* B.

We have already seen that the idea of an element being neither *before* nor *after* another element, and yet distinct from it, involves no logical absurdity, and so if we give up the attempt to identify the instant B with any instant at P we get a logically consistent view of things.

Thus according to the view here adopted *there is no identity of instants at different places at all.*

We may express the idea in this form: *the present instant, properly speaking, does not extend beyond here.*

Thus there are instants at a distance *before* the present instant and *after* it, and also instants neither *before* nor *after* it, but such instants are to be regarded as being all quite distinct from the present instant here.

Thus, according to the view here adopted, *the only really simultaneous events are events which occur at the same place.*

The theory which we desire to expound with regard to time and space may now briefly be described as follows:

Taking the above view of instants and the relations of *before* and *after*, we express in terms of these relations the conditions that the set of instants should have a *conical order* of a certain type. We then find that we have got a description not only of time but also of space such as that with which we are already familiar.

In fact we may be said to analyze spacial relations in terms of the time relations of before and after.

In first approaching this subject it is a great assistance to have some concrete way of representing the facts to our minds even though such representation may make use of some of the conceptions which we are trying to analyze.

In doing so one must remember however that the justification of our theory lies in the logical procedure and not in the representation.

Thus in trying to convey a general idea of what we are doing we shall find it both convenient and suggestive to make use of our mental images of cones, in the way already described, in order to picture what we mean by *conical order.*

The idea of *conical order* is not at all dependent on this representation, but is built up by a rather lengthy piece of reasoning from the relations of *before* and *after.*

The representation by means of cones may be compared to the rough scaffolding used in the erection of a building, which is removed when the building is complete and its component parts in position.

We must, however, be certain that the building is not supported by the scaffolding, or it will not be able to stand alone.

In order to make sure of this in our theory, great care has to be taken, and, for details on this matter, I must refer readers to my larger work.

Moreover, the representation by means of cones gives only a three-dimensional conical order, whereas the conical order with which we are really concerned is a four-dimensional one.

The representation also introduces a sort of distortion, but this need not trouble us when we deal only with descriptive features.

Now in ordinary mathematical physics we are accustomed to localize an instantaneous event by means of four numbers x, y, z, t. Of these numbers x, y and z are called spacial coordinates while t is referred to as the "time."

But now having come to regard all instants at different places as distinct, we regard these four numbers as really representing four coordinates of an instant.

The coordinate t, however, has different *before* and *after* relations from those associated with the other three coordinates x, y, z, which are made clear by the conception of *conical order*.

In order to avoid confusion therefore, we shall speak of the former not as "time" but as a t coordinate.

We are not yet, however, in a position to introduce coordinates except for "scaffolding" purposes.

Neither again are we at liberty to speak of "velocity" except for scaffolding purposes until we have defined the meaning of the word.

Moreover, in the actual proof of theorems, we must not employ the ideas of equality of lengths or angles until these ideas are seen to be definable in terms of *before* and *after* relations.

We may, however, make use of such terms in the "scaffolding" which is mere poetry and rhetoric.

Let us therefore first consider this pictorial representation in which we have to confine ourselves to three coordinates instead of four, which we shall take to be x, y and t, and shall regard as rectangular.

Now by taking suitable units we may regard the "velocity of light" as unity and under these circumstances if we imagine a flash of light starting from the position $x = a$, $y = b$, $t = c$, the rays of light would be represented by the generators of the upper half of the cone whose equation is

$$(x - a)^2 + (y - b)^2 - (t - c)^2 = 0,$$

which we take as the α cone corresponding to (a, b, c).

The lower portion of this same locus constitutes the β cone of (a, b, c).

The point (a, b, c) itself is regarded as belonging to both the α and β cones.

The successive positions of a material particle would be represented by some line straight or curved, but since it appears that a particle of matter never quite attains to the "velocity of light" the tangent to this curve would make an angle with the axis of t which is always less than 45°: the semi-vertical angle of the cones.

The successive positions of a particle which remains at rest with respect to the system of axes would be represented by a straight line parallel to, or coincident with, the axis of t.

The successive positions of a particle which remains in uniform motion with respect to the system of axes would also be represented by a straight line, but one inclined to the axis of t.

The successive positions of an accelerated particle would be represented by a curved line.

The set of instants of which any one individual is *directly conscious* would also be represented by some line straight or curved, whose tangent always makes an angle with the axis of t less than the semi-vertical angle of the cones.

We thus see that for the set of instants of which any one individual is directly conscious, or the set of instants which any one particle occupies, we can assert that: if an instant A be neither *before* nor *after* an instant B, the instants A and B are identical.

We cannot, however, assert this of the instants of which two individuals are directly conscious, or which two distinct and separate particles occupy.

It may be that an instant of which I am directly conscious is neither *before* nor *after* some instant of which you are directly conscious, but they are not identical, and our illustration shows that this involves no logical contradiction.

It is to be noted, however, that if A and B be two distinct instants, one of which is neither *before* nor *after* the other, then since the α cones intersect and also the β cones, there are instants which are *after* both A and B and also instants which are *before* both A and B, so that we may both speak of to-morrow or of yesterday, though strictly speaking we have no common present.

Thus instead of regarding ourselves as, so to speak, swimming along in an ocean of space (as we usually do), we are to think of ourselves rather as swimming along in an ocean of time, while *spacial relations are to be regarded as the manifestation of the fact that the elements of time form a system in conical order: a conception which may be analyzed in terms of the relations of after and before.*

Having thus given a sort of vista of the promised land, we must next give some account of the rather toilsome journey entailed in getting there.

CONICAL ORDER

In building up the idea of *conical order* in terms of the relations of *before* and *after*, we make use of certain postulates involving these relations.

These postulates do not involve any idea of measurement but are of a purely descriptive character.

If there are any physical facts corresponding to the postulates, then we shall be able to describe such facts in terms of the relations of *before* and *after*.

In the formal development of the theory we shall speak of elements instead of instants and of α and β sub-sets instead of α and β cones.

The α and β sub-sets have, however, yet to be defined.

When we wish to form a concrete picture of these we shall frequently make use of the cones, since by doing so we get suggestions as to suitable postulates, and also as to methods of procedure to prove theorems.

However, we supposed the cones to have their axes parallel and to have equal vertical angles, and neither the idea of *cone*, of *parallel*, of *axis*, of *angle*, or of *equal* have been analyzed in terms of *before* and *after*, and must therefore be excluded in defining α and β sub-sets.

The relations of *before* and *after* being converse asymmetrical relations, either may be defined in terms of the other.

It is a matter of indifference which we take as fundamental, but in the following account we shall start with the relation of *after*.

Most of the postulates consist of two parts marked (*a*) and (*b*) in which the relations *before* and *after* are interchanged.

In some, however, such as those numbered I, III and IV, the one part follows from the other as a direct consequence of the mutual relations of *after* and *before*, while in others such as V, these relations are involved symmetrically.

We shall now proceed with an account of the formal development of the theory.

We shall suppose that we have a set of elements and that certain of these elements stand in a relation to certain other elements of the set which we denote by saying that one element is *after* another.

The first four postulates are the equivalents of the first four characteristics which we observed as belonging to the set of instants of which any one individual is directly conscious.

We shall re-state them as follows:

POSTULATE I. **If an element B be after an element A, then the element A is not after the element B.**

Definition. If an element B be *after* an element A, then the element A will be said to be *before* the element B.

POSTULATE II. (*a*) **If A be any element, there is at least one element which is after A.**

(*b*) **If A be any element, there is at least one element which is before A.**

POSTULATE III. **If an element B be after an element A, and if an element C be after the element B, the element C is after the element A.**

POSTULATE IV. **If an element B be after an element A, there is at least one element which is both after A and before B.**

The next postulate is one to admit of the existence of pairs of instants, of which not more than one of a pair can be in the direct consciousness of any single individual; or in other words, of the existence of instants which, in respect of any one individual, are, as we say, *elsewhere*. It is as follows:

POSTULATE V. **If A be any element, there is at least one other element distinct from A, which is neither before nor after A.**

In our illustration by means of cones an element distinct from A, and neither *before* nor *after* it, is represented by a point outside both the α and β cones of A.

If B be such a point then as we have also seen, the α cones of A and B intersect as do the β cones of A and B.

Now we wish to express in terms of *before* and *after* a characteristic property of a point of intersection.

If we take X as a point on the locus of intersection of the α cones of A and B and we take the plane through A, B and X, we observe in the first place that X is *after* both A and B.

Further X is *before* all other points in the plane which are *after* both A and B.

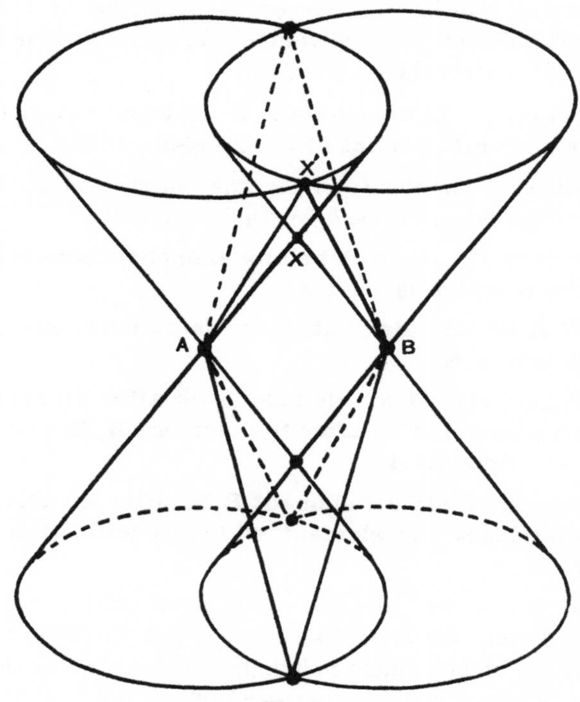

Fig. 3

This is not, however, the case if we go outside this plane, for if we take a second point X' on the locus of intersection, then X is not *before* X', although X' is *after* both A and B.

We can, however, easily see that though we are not at liberty to say that X is *before*, we can assert that X is *not after* any other point which is *after* both A and B, and this gives us the property we have been searching for.

Thus we may express our next postulate as follows:

Postulate VI. (*a*) **If A and B be two distinct elements, one of which is neither before nor after the other, there is at least one element which is after both A and B, but is not after any other element which is after both A and B.**

(*b*) **If A and B be two distinct elements, one of which is neither after nor before the other, there is at least one element which is before both A and B, but is not before any other element which is before both A and B.**

This last postulate, although somewhat complicated in form, is extremely important as it enables us to give a definition of the α and β sub-sets.

In fact reverting to our illustration, a point X which bears the relation postulated in Post. VI (*a*) to two points A and B, one of which is neither *before* nor *after* the other, must lie in the α cones of both A and B.

Further, if X lies in the α cone of A and is distinct from it, there exist points, such as B, which in general we may call Y.

As we wish to include A itself in the α cone or sub-set, we mention it explicitly.

Thus we get the following definitions of the sub-sets:

Definition. (*a*) If A be any element of the set, then an element X will be said to be a member of the α sub-set of A provided X is either identical with A, or else provided there exists at least one element Y distinct from A and neither *before* nor *after* A and such that X is *after* both A and Y but is not *after* any other element which is *after* both A and Y.

(*b*) If A be any element of the set, then an element X will be said to be a member of the β sub-set of A provided X is either identical with A, or else provided there exists at least one element Y distinct from A and neither *after* nor *before* A and such that X is *before* both A and Y but is not *before* any other element which is *before* both A and Y.

We must next express some further properties of the α and β sub-sets which obviously hold in the case of our cones.

We shall denote by α_1 and β_1 the sub-sets corresponding to an element A_1, and by α_2 and β_2 those corresponding to an element A_2, etc.

Postulate VII. (*a*) **If $\mathbf{A}_1$ and $\mathbf{A}_2$ be elements and if $\mathbf{A}_2$ be a member of α_1 then $\mathbf{A}_1$ is a member of β_2.**

(*b*) **If $\mathbf{A}_1$ and $\mathbf{A}_2$ be elements and if $\mathbf{A}_2$ be a member of β_1 then $\mathbf{A}_1$ is a member of α_2.**

Postulate VIII. (*a*) **If $\mathbf{A}_1$ be any element and $\mathbf{A}_2$ be any other element in α_1, there is at least one other element distinct from $\mathbf{A}_2$ which is a member both of α_1 and of α_2.**

(*b*) **If $\mathbf{A}_1$ be any element and $\mathbf{A}_2$ be any other element in β_1, there is at least one other element distinct from $\mathbf{A}_2$ which is a member both of β_1 and of β_2.**

By the help of the above postulates we can prove a number of theorems.

Thus our first theorem is that: if A_1 be any element and A_2 be any other element in α_1, then any element A_3 which is both *after* A_1 and *before* A_2 must be a member both of α_1 and β_2.

This is easily seen to hold in the case of the cones.

We can also show that there are elements in the α sub-set of any element which are distinct and neither *before* nor *after* one another; and similarly in the β sub-set.

The next step is to define what corresponds to a generator of a complete cone, or α and β taken together.

Now reverting to our illustration, we notice that if one point lies on the α cone of another and is distinct from it, then the two cones touch along a generator.

We accordingly introduce the following:

Definition. If A_1 be any element and A_2 be any other element in α_1, the *optical line* A_1A_2 is defined as the aggregate of all elements which lie either

(1) both in α_1 and α_2,

or (2) both in α_1 and β_2,

or (3) both in β_1 and β_2.

Before we can prove the chief properties of what we have called an optical line it is necessary to introduce another postulate.

Taking our usual illustration, the thing which represents an *optical line* is a generator of the combined α and β cones.

Now if we consider any such generator there are points which, while not being points of the generator itself, are *before* points of it and similarly there are points which are not points of the generator but are *after* points of it.

We are able from the postulates already given to prove the existence of elements having corresponding properties.

Now if we consider a point which is not a point of a given generator but is *before* some point of it, it is easy to see that the α cone of the first point has one single point in common with the generator.

We have also a corresponding result if we consider a point which is not a point of the generator but *after* some point of it, in which case the β cone of the first point has one single point in common with the generator.

We accordingly introduce the following postulate:

POSTULATE IX. (*a*) **If a be an optical line and $\mathbf{A}_1$ be any element which is not in the optical line but before some element of it, there is one single element which is an element both of the optical line a and the sub-set α_1.**

(*b*) **If a be an optical line and $\mathbf{A}_1$ be any element which is not in the optical line but after some element of it, there is one single element which is an element both of the optical line a and the sub-set β_1.**

We can now prove a number of theorems which are important in the logical development, such as the existence of elements having certain specific properties.

We can also prove that of any two elements of an optical line the one lies in the α sub-set of the other and is therefore *after* it.

It can also be shown that there are an infinite number of elements in an optical line and that they are in a linear order as distinguished from a conical one.

It can also be proved that any two elements of an optical line determine that optical line.

Further, we have the interesting result that if an element A_1 be *before* an element of an optical line and also *after* an element of it, then A_1 is itself an element of the optical line.

Having thus defined an optical line and proved some of its chief properties, the next obvious thing to do is to try to define the representatives of lines which are not generators of α and β cones, but this is not so easy. The method by which I succeeded in doing this was by taking the intersections of two planes of a certain type.

Making use of our illustration once more, we notice that if we take the combined α and β cones of any point and take a straight line passing through the vertex, such straight line may either (i) be a generator of the cone (corresponding to an optical line); or (ii) may fall inside the cone; or (iii) may fall outside the cone.

Again, if we take a plane through the vertex it may either (i) be a tangent plane to the cone; or (ii) may cut the cone in two generators; or (iii) may have no real point in common with the cone except the vertex.

Only the second type of plane contains all three types of line, but the method of defining such a plane offers some difficulties.

We may note, however, that through any point of such a plane there are two and only two lines of the type of what we have called optical lines which lie entirely in the plane. In fact there are two distinct parallel sets of these lines lying in the plane.

Now the method employed in defining a plane in most work on the foundations of geometry is to take a triangle and to define the plane as the aggregate of all points of all lines which intersect two side lines of the triangle in distinct points; or some equivalent of this.

This method is not, however, open to us, since the only lines we have defined cannot form a triangle.

Another method which suggests itself is to take two parallel lines and to define the plane as the aggregate of all points of all of a system of lines which intersect the first two.

This seems more hopeful, but the difficulty remains of defining the parallelism of the two lines to begin with.

The ordinary definition of parallel lines implies the prior notion of a plane in which the lines lie.

We thus seem to be up against a formidable difficulty.

But now we notice that only planes of types (i) and (ii) contain lines of the character of optical lines.

A plane of type (i) contains only one set of parallel lines of this character, while a plane of type (ii) contains two sets of such parallel lines.

Now a little consideration will show us that if a and b be two parallel lines of the type corresponding to optical lines lying in a plane of type (i), then no point of a is either *before* or *after* any point of b.

If the lines a and b lie in a plane of type (ii), then either each point of a is *after* a point of b, or else each point of a is *before* a point of b. If finally we take two lines a and b of this type which are oblique to one another and do not intersect, it is not difficult to see that there are some points of a which are *after* points of b and others which are *before* points of b and these two sets are separated by a single point of a which is neither *before* nor *after* any point of b.

These considerations give us the clue to a new postulate as follows:

POSTULATE X. (a) **If a be an optical line and if A be any element not in the optical line but before some element of it, there is one single optical line containing A and such that each element of it is before an element of a.**

(b) **If a be an optical line and if A be any element not in the optical line but after some element of it, there is one single optical line containing A and such that each element of it is after an element of a.**

We can easily show that if each element of one optical line be *before* an element of another distinct optical line, the two optical lines cannot have an element in common.

Similarly, if each element of the one be *after* an element of the other.

We can also prove that if each element of an optical line a be *before* an element of another optical line b, then through each element of a there is one single optical line which intersects b; and similarly if each element of a be *after* an element of b.

We are not, however, as yet in the position to carry out the plan which we outlined for defining the types of line other than optical lines.

It will be remembered that it was proposed to define a certain type of plane and to define these other types of line by means of the intersection of two such planes.

To have two planes intersecting it would be necessary to have more than two dimensions, and all the postulates which we have hitherto introduced may be represented in one plane.

It is easy to introduce a postulate which gives *at least* a three-dimensional character to our system, and this might have been done by a slight alteration of Postulate VI (a) and (b).

If we substitute in it for the words "*there is at least one element which*" the words: *there are at least two elements either of which*, we should have had what we wanted; but leaving Postulate VI in its original form we can express the same thing by a new postulate as follows:

POSTULATE XI. (*a*) **If A_1 and A_2 be two distinct elements, one of which is neither before nor after the other, and X be an element which is a member both of α_1 and α_2, then there is at least one other element distinct from X which is a member both of α_1 and α_2.**

(*b*) **If A_1 and A_2 be two distinct elements, one of which is neither after nor before the other, and X be an element which is a member both of β_1 and β_2, then there is at least one other element distinct from X which is a member both of β_1 and β_2.**

It is easily seen that if A_1 and A_2 be two distinct elements such that the one is neither *before* nor *after* the other and if X and X' be two distinct elements lying both in α_1 and in α_2, then X is neither *before* nor *after* X'.

Similarly if X and X' lie in both β_1 and β_2.

It is not difficult to prove now that if A be any element there are at least three distinct optical lines containing A.

It will be observed, however, from our illustration that not more than two of the lines corresponding to optical lines which pass through any point lie in any one plane.

Now we have already observed in our illustration that if a and b be two of the lines corresponding to optical lines which neither intersect nor are parallel, there is one single point of a which is neither *before* nor *after* any point of b, and we now wish to make use of the corresponding property of optical lines in order to investigate their parallelism. We accordingly introduce another postulate as follows:

POSTULATE XII. (*a*) **If a be an optical line and A be any element not in the optical line but before some element of it, then each optical line through A, except the one which intersects a and the one of which each element is before an element of a, has one single element which is neither before nor after any element of a.**

(*b*) **If *a* be an optical line and A be any element not in the optical line but after some element of it, then each optical line through A, except the one which intersects *a* and the one of which each element is after an element of *a*, has one single element which is neither after nor before any element of *a*.**

We are now able to prove that: if each element of an optical line *a* be *after* an element of a distinct optical line *b*, then each element of *b* is *before* an element of *a* and *vice versa*.

This is not obvious, as might perhaps at first sight appear.

We can also prove now that: if *a* be an optical line and if A_1 be any element which is neither *before* nor *after* any element of *a*, there is *one single optical* line containing A_1 and such that no element of it is either *before* or *after* any element of *a*.

We have already seen how this is represented in our illustration.

We can now define the parallelism of optical lines as follows:

Definitions. An optical line *a* will be said to be parallel to a second distinct optical line *b* when either

(1) each element of *a* is *after* an element of *b*,

or (2) each element of *a* is *before* an element of *b*,

or (3) no element of *a* is either *before* or *after* any element of *b*.

In case (1) *a* will be said to be an after-parallel of *b*.

In case (2) *a* will be said to be a before-parallel of *b*.

In case (3) *a* will be said to be a neutral-parallel of *b*.

We can show that if *a* be parallel to *b*, then *b* is parallel to *a*.

Also we can show that if *a* be an optical line and *A* be any element not in the optical line, then there is one single optical line parallel to *a* and containing *A*.

Further, we are able to prove a number of theorems which combined together give us the general result that:

If two distinct optical lines *a* and *b* are each parallel to a third optical line *c*, then the optical lines *a* and *b* are parallel one to another.

We may now give the following definition:

Definition. If *a* and *b* be any pair of distinct optical lines one of which is an after-parallel of the other, then the aggregate of all elements of all optical lines which intersect both *a* and *b* will be called an *acceleration plane*.

As we have already seen in our illustration what we have called an acceleration plane is represented by a plane through the vertex of a cone of the system intersecting the cone in two distinct generators.

We have given it the name *acceleration plane* because it might be supposed to be determined by the acceleration of a particle, but for the present we shall simply take it as a name.

We can easily prove that there are an infinite number of acceleration planes which contain any given optical line.

We can now introduce a new postulate.

POSTULATE XIII. **If two distinct acceleration planes have two elements in common, then any other acceleration plane containing these two elements contains all elements common to the two first mentioned acceleration planes.**

We can now prove that if a and b be two distinct optical lines and if a be an after-parallel of b, then if c and d be two other distinct optical lines intersecting both a and b, one of the optical lines c, d is an after-parallel of the other.

We can also show that an acceleration plane contains two and only two optical lines which pass through any element of it and these form two parallel systems.

Further, if an acceleration plane contain an optical line a and an element A_1 which does not lie in the optical line, then A_1 is either *before* or *after* an element of a.

We can, moreover, show that there are an infinite number of acceleration planes containing any pair of elements, whether the one be *after* the other or neither *before* nor *after* it.

We can also show that if two or more distinct acceleration planes contain an optical line, there is no other element which they have in common which does not lie in the optical line.

We may now introduce the following definitions:

Definitions. If two acceleration planes contain two elements in common, then the aggregate of all elements common to the two acceleration planes will be called a *general line.*

If two acceleration planes contain two elements in common, of which one is *after* the other but does not lie in the same optical line with it, then the aggregate of all elements common to the two acceleration planes will be called an *inertia line.*

If two acceleration planes contain two elements in common, of which one is neither *before* nor *after* the other, then the aggregate of all elements common to the two acceleration planes will be called a *separation line*.

The name *inertia line* is employed because it appears to represent the set of instants which an unaccelerated particle occupies; while the name *separation line* is used because particles which occupy distinct elements of such a line would be separated, as we say, "in space."

In the formal development of our subject, however, we shall simply regard these as names.

If we take our usual illustration and consider a complete cone, then an inertia line would be represented by a straight line passing through the vertex and lying inside the cone, while a separation line would be represented by one passing through the vertex but lying outside the cone.

In order to investigate the properties of an inertia line we introduce a fresh postulate whose representation in our illustration is obvious.

POSTULATE XIV. (*a*) **If *a* be any inertia line and $\mathbf{A}_1$ be any element of the set, then there is one single element common to the inertia line *a* and the sub-set α_1.**

(*b*) **If *a* be any inertia line and $\mathbf{A}_1$ be any element of the set, then there is one single element common to the inertia line *a* and the sub-set β_1.**

This should be compared with Post. IX.

It is to be observed that any element of the set is *after* certain elements of any inertia line and *before* certain others, whereas if any element be *after* one element of an optical line and *before* another element of it, it must itself be an element of the optical line. Moreover, it follows from Post. XII that there are elements of the set which are neither *before* nor *after* any element of a given optical line.

We are now in a position to prove a number of the characteristic properties of an inertia line.

Thus we can show that an inertia line in any acceleration plane has one single element in common with each optical line in the acceleration plane.

We can also show that of any two distinct elements of an inertia line one is *after* the other and so no inertia line can be a separation line.

We can in fact show that the elements of an inertia line have the whole five characteristic properties which we mentioned as belonging to the set of instants of which any one individual is directly conscious.

This does not mean, however, that the latter set of instants are confined to an inertia line.

In order to investigate the properties of a separation line we introduce the following postulate:

POSTULATE XV. **If two general lines, one of which is a separation line and the other is not, lie in the same acceleration plane, then they have an element in common.**

We further make the following definition:

Definition. An element in an acceleration plane will be said to be *between* a pair of parallel optical lines in the acceleration plane if it be *after* an element of the one optical line and *before* an element of the other and does not lie in either optical line.

We can now show that if A_1 and A_2 be any two distinct elements of a separation line, there is at least one other element of the separation line which lies between a pair of parallel optical lines through A_1 and A_2 respectively in an acceleration plane containing the separation line.

We can also show that any such element lies between the other pair of parallel optical lines in the acceleration plane which pass through A_1 and A_2.

In fact by making use of parallel optical lines we can assign an order to the elements of a separation line, in so far as a particular acceleration plane is concerned, but we are not yet in a position to show that this order is independent of the particular acceleration plane in which the separation line is regarded as lying.

Another interesting result which we may mention is that if A be an element of an optical line a and if B be an element which is neither *before* nor *after* any element of a, then no element of the separation line AB, with the exception of A, is either *before* or *after* any element of a.

It is clear that in this case the optical line a and the element B cannot lie in one acceleration plane. In fact they lie in another type of plane which is represented in our illustration by a tangent plane to a cone.

Now if A_1, A_2 and A_3 be three distinct elements which do not all lie in one general line, this last result shows that they either may or may not lie in one acceleration plane.

If they do lie in one acceleration plane they must determine the acceleration plane, since should they lie in a second one they would have to lie in one general line, contrary to hypothesis.

It is possible at this stage to give certain criteria by which we can say whether or no a set of three elements does lie in one acceleration plane.

These will be found in my larger work.

We can now define the parallelism of acceleration planes.

If we call the optical lines in an acceleration plane the *generators* of the acceleration plane this definition is as follows:

Definition. If an acceleration plane have its two sets of generators respectively parallel to the two sets of generators of another distinct acceleration plane, then the two acceleration planes will be said to be *parallel* to one another.

It is easy to see that if P be an acceleration plane and A be any element outside it, then there is one single acceleration plane containing A and parallel to P, and further this acceleration plane can have no element in common with P.

Further, two distinct acceleration planes which are parallel to the same acceleration plane are parallel to one another.

There is no distinction of different sorts of parallelism of acceleration planes as there is for optical lines.

We can now prove that: if an acceleration plane P have one element in common with each of a pair of parallel acceleration planes Q and R then, if P have a second element in common with Q it has also a second element in common with R.

Now we have already noticed that if two distinct optical lines intersect a pair of optical lines, one of which is an after-parallel of the other, then of the two first-mentioned optical lines one is an after-parallel of the other.

We shall accordingly state the following definition:

Definition. If two distinct optical lines intersect a pair of optical lines, one of which is an after-parallel of the other, then the four optical lines will be said to form an *optical parallelogram.*

It is evident that an optical parallelogram lies in an acceleration plane.

We can give obvious definitions to *corners*, *opposite*, or *adjacent*.

We can also give obvious definitions to *diagonal lines* of an optical parallelogram, and it is easy to see that one of these must be an inertia line and the other a separation line.

We shall now give a new postulate which gives a relation between certain inertia lines and certain separation lines.

POSTULATE XVI. **If two optical parallelograms lie in the same acceleration plane, then if their diagonal lines of one kind do not intersect, their diagonal lines of the other kind do not intersect.**

It is now possible to show that if a be any general line in an acceleration plane P and A_1 be any element of the acceleration plane which is not in the general line, then there is one single general line through A_1 in the acceleration plane which does not intersect a, and further, this general line must be of the same type as a.

We can also show that if two acceleration planes P and Q have a general line a in common, and if A_1 be any element which does not lie either in P or Q, then the acceleration planes through A_1 parallel to P and Q, respectively, have a general line in common.

This last result enables us to give a definition of the parallelism of a pair of general lines whether these lie in one acceleration plane or not.

It may be expressed as follows:

Definition. If a be a general line, and A be any element which does not lie in it, and if two acceleration planes R and S through A are parallel respectively to two others P and Q containing a, then the general line which R and S have in common is said to be parallel to a.

It is easy to see that this covers the case of the parallelism of optical lines which is the only case of the parallelism of general lines which we had hitherto defined.

We are able to show now that: if a be a general line, and A_1 be any element which does not lie in it, there is one single general line containing A_1 and parallel to a.

We can also show that if two distinct general lines are each parallel to a third, then they are parallel to one another.

If a and b be any pair of parallel general lines, it is easy to see that they must be general lines of the same kind, for we have found in the course of our work that two parallel general lines in one acceleration plane must be of the same kind, and by two applications of this result it follows that if a and b do not lie in one acceleration plane they must also be of the same kind.

A number of other theorems concerning parallel general lines may now be proved which are analogous to corresponding theorems about parallel straight lines in ordinary geometry.

These are important in the strict logical development of our subject, but as we are here only concerned with giving an outline of the course of procedure, I must again refer readers to my larger work.

Definition. The element of intersection of the diagonal lines of an optical parallelogram will be called the *centre* of the optical parallelogram.

We can now prove that: if two distinct elements A and O be taken in an inertia or separation line in a given acceleration plane, then there is one single optical parallelogram in the acceleration plane having O as centre and A as one of its corners.

We can also give demonstrations of some other very important theorems concerning optical parallelograms.

We can show in the first place that: if two optical parallelograms have two opposite corners in common, then they have a common centre.

For the optical parallelograms to be distinct, it is clear that they must lie in different acceleration planes.

Again, we can show that: if two optical parallelograms have two adjacent corners in common, then optical lines through the centres of the optical parallelograms and intersecting their common side line intersect it in the same element.

Here the two optical parallelograms either may, or may not, lie in the same acceleration plane.

We can now introduce the following definitions:

Definition. If A and B be two distinct elements lying in an inertia line or in a separation line, then the centre of an optical parallelogram of which A and B are a pair of opposite corners will be spoken of as the *mean* of the elements A and B.

Definition. If A and B be two distinct elements lying in an optical line, then an optical line through the centre of an optical parallelogram of which A and B are a pair of adjacent corners and intersecting the optical line AB, intersects it in an element which will be spoken of as the *mean* of the elements A and B.

The above-mentioned theorems show that the mean of the

elements A and B is a definite element independent of the particular optical parallelogram used to define it. This marks the first stage on the way to introducing the ideas of equality of lengths and of measurement.

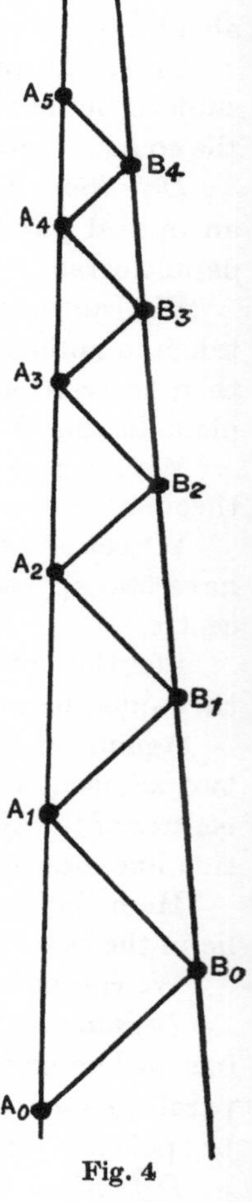

Fig. 4

We have now to introduce a new postulate which bears a sort of analogy to the well-known *axiom of Archimedes*, but which, unlike the latter, does not contain any reference to congruence.

Before doing so, however, it is necessary to go into certain points in connexion with it.

If a and b be any two distinct inertia lines, and A_0 be any element in a which is not an element of intersection with b, then from Post. XIV (a) it follows that there is one single element common to the inertia line b and the α sub-set of A_0.

Call this element B_0.

Then B_0 is distinct from A_0 and cannot be an element of intersection of the two inertia lines, for if it were A_0 and B_0 would lie both in an inertia line and an optical line which is impossible.

Further, there cannot be an element of intersection of the inertia lines lying *after* A_0 and *before* B_0, for then, as was pointed out on page 21, such element would require to lie in the optical line A_0B_0 and so again we should have two elements lying both in an optical line and an inertia line which is impossible.

Thus any element of intersection of the two inertia lines, if such an element exists, must lie either *before* A_0 or *after* B_0.

Again, from Post. XIV (a) it follows that there is one single element, say A_1, common to the inertia line a and the α sub-set of B_0, and again, A_1 cannot be an element of intersection of the inertia lines.

Further, any such element, if it exists, must lie either *before* A_0 or *after* A_1.

Proceeding again in the same way, there is one single element, say B_1, common to the inertia line b and the α sub-set of A_1, and one single element A_2 common to the inertia line a and the α sub-set of B_1, and so on.

We thus get an infinite series of elements A_0, A_1, A_2, A_3 ... in the inertia line a and another infinite series of elements B_0, B_1, B_2, B_3 ... in the inertia line b.

An element of intersection of the two inertia lines, if such an element exists, must lie either *before* A_0 or *after* A_n, where n is any finite integer whatever.

This process will be spoken of as *taking steps along the inertia line a with respect to the inertia line b.*

The passing from A_0 to A_1 is the first step, the passing from A_1 to A_2 the second, and so on.

If X be an element in a which is *after* A_0 and *before* A_n but not *before* A_{n-1}, then the element X will be said to be *surpassed* from A_0 in n steps taken along a with respect to b.

If C be an element of intersection of the two inertia lines, and if C be *after* A_0, it is evident from what we have said that C *cannot be surpassed from* A_0 *in any finite number of steps.*

We can now introduce our new postulate.

POSTULATE XVII. **If A_0 and A_x be two elements of an inertia line a such that A_x is after A_0, and if b be a second inertia line which does not intersect a either in A_0, A_x or any element both after A_0 and before A_x, then A_x may be surpassed in a finite number of steps taken from A_0 along a with respect to b.**

It follows directly from Post. XVII that if the two inertia lines *do not intersect at all*, then A_x may *always* be surpassed in a finite number of steps.

There is also a (b) form of this postulate, but it is not independent, and in fact is proved in Theorem 64 of my larger work.

The primary use to which we put this postulate is to prove certain theorems which entail repeated constructions of some particular type.

Without the postulate we should have no guarantee that certain elements came within the scope of our proofs.

Making use of results obtained in this way, we are now able to prove the following important theorem:

If A, B and C be three elements in a separation line, and if

B be between a pair of parallel optical lines through *A* and *C* in an acceleration plane containing the separation line, then *B* is also between a pair of parallel optical lines through *A* and *C* in any other acceleration plane containing the separation line.

This is a result which clearly holds in our ordinary illustration, but hitherto we had not been able to demonstrate it from our postulates.

The importance of the theorem lies in the fact that we can now assign a definite order to the elements of a separation line, which is independent of a particular acceleration plane.

We accordingly introduce the following definition:

Definition. If three distinct elements lie in a general line, and if one of them lies between a pair of parallel optical lines through the other two in an acceleration plane containing the general line, then the element which is between the parallel optical lines will be said to be ***linearly between*** the other two elements.

The above definition is so framed as to apply to all three types of general line and is therefore more complicated than it need be if we were dealing only with optical or inertia lines.

For the case of elements lying in either of these types of general line, one element is linearly between two other elements if it be *after* the one and *before* the other.

In the case of elements lying in a separation line, however, no one is either *before* or *after* another, and so we have to fall back on our definition involving parallel optical lines.

The distinction between the three cases is interesting, and is made clear if we make use of our usual illustration.

Thus if three elements *A*, *B* and *C* lie in a general line *a*, and if *B* be linearly between *A* and *C*, then in case *a* be an inertia line we must either have *B after A*, and *C after B*, or else *B after C* and *A after B*, and similarly when *a* is an optical line.

If *a* be an inertia line, and *B* be *after A*, and *C after B*, then *B* will be *before* elements of both optical lines through *C*, and *after* elements of both optical lines through *A* in any acceleration plane containing *a*.

If *a* be an optical line, and *B* be *after A*, and *C after B*, then, apart from *a* itself, there is only one optical line through any element of *a* in any acceleration plane containing *a*, and so we should have *B before* an element of the optical line through *C* and *after* an element of the parallel optical line through *A*.

If a be a separation line, however, we should have B *before* an element of one of the optical lines through C, and *after* an element of the parallel optical line through A, and also *after* an element of the second optical line through C, and *before* an element of the parallel optical line through A in any acceleration plane containing a.

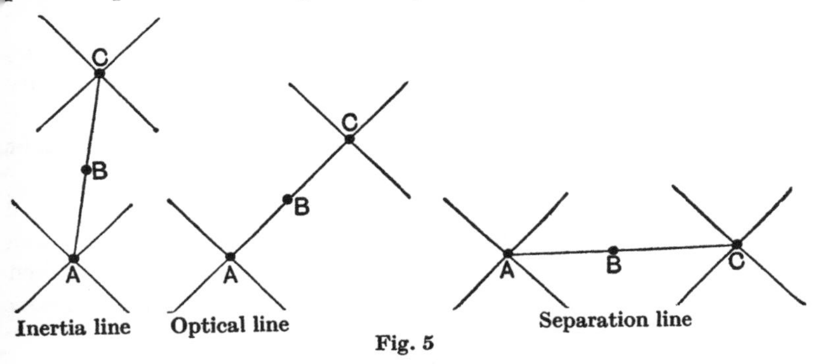

Fig. 5

Now Peano has given a number of axioms of the straight line in ordinary geometry in terms of the relation of *between* and the corresponding properties may now be shown to hold for the general line in our geometry in terms of the relation of *linearly between* as we have defined it.

We can also prove two theorems which correspond to two axioms given by Peano for points in a plane.

These theorems are as follows:

If A, B and C be three elements in an acceleration plane which do not all lie in one general line, and if D be an element linearly between A and B, while E is an element linearly between B and C, there exists an element which lies both linearly between A and E and linearly between C and D.

If A, B and C be three elements in an acceleration plane which do not all lie in one general line, and if D be an element linearly between A and B, while F is an element linearly between C and D, there exists an element, say E, which is linearly between B and C, and such that F is linearly between A and E.

The proofs of these theorems require the consideration of a number of different cases and are therefore somewhat tedious.

It is to be noted that these theorems have as yet only been proved for elements in an acceleration plane.

In addition to the analogues of the various axioms of Peano

concerning the ordering of points in a line or plane, we have also seen that if a be any general line in an acceleration plane, and A be any element of the acceleration plane which does not lie in a, then there is one single general line passing through A and lying in the acceleration plane which does not intersect a.

This corresponds to the Euclidean axiom of parallels.

An acceleration plane, however, differs from a Euclidean plane, since there are three types of general line in the former and only one type of straight line in the latter.

The congruence properties of the two are also quite different as will be seen later.

We shall now introduce the following definition:

Definition. If two parallel general lines in an acceleration plane be both intersected by another pair of parallel general lines, then the four general lines will be said to form a *general parallelogram in the acceleration plane.*

It should be noted that we shall afterwards extend the term general parallelogram to figures in other types of plane.

We can now prove that if we have a general parallelogram in an acceleration plane, then:

(1) The two diagonal lines intersect in an element which is the mean of either pair of opposite corners.

(2) A general line through the element of intersection of the diagonal lines and parallel to either pair of side lines, intersects either of the other side lines in an element which is the mean of the pair of corners through which that side line passes.

We can also prove the theorem that: if A, B and C be three elements in an acceleration plane which do not all lie in one general line, and if D be the mean of A and B, then a general line through D parallel to BC intersects AC in an element which is the mean of A and C.

Again, we can show that if A_0 and A_n be two distinct elements in a general line a, we can always find $n-1$ elements $A_1, A_2, \dots A_{n-1}$ in a (where $n-1$ is any integer), such that:

A_1 is the mean of A_0 and A_2,

A_2 is the mean of A_1 and A_3,

....

....

A_{n-1} is the mean of A_{n-2} and A_n.

We have now to take up the study of the type of plane which in our illustration is represented by a tangent plane to a complete cone.

The first step consists in proving certain theorems in connexion with neutral-parallel optical lines. Thus we can prove the following:

If A be any element in an optical line a, and A' be any element in a neutral-parallel optical line a', then if B be a second element in a the general line through B parallel to AA' intersects a' in an element, say B'.

Further, if B be *after* A, then B' is *after* A', while if B be *before* A then B' is *before* A'.

Again, we can show that under the above conditions there is only one general line through B and intersecting a' which does not intersect the general line AA': namely, the parallel to AA'.

Again, if a and b be two neutral-parallel optical lines, and if one general line intersects a in A and b in B, while a second general line intersects a in A' and b in B', then an optical line through any element of AB and parallel to a or b intersects $A'B'$.

We can now show that if a and b be two neutral-parallel optical lines, and if c and d be any two non-parallel separation lines which intersect both a and b, then the aggregate consisting of all the elements in c and in all separation lines intersecting a and b which are parallel to c must be identical with the aggregate consisting of all the elements in d and in all separation lines intersecting a and b which are parallel to d.

This prepares the way for the following definition:

Definition. The aggregate of all elements of all mutually parallel separation lines which intersect two neutral-parallel optical lines will be called an *optical plane*.

It is evident that through any element of an optical plane there is one single optical line lying in the optical plane.

In analogy with the case of an acceleration plane, an optical line which lies in an optical plane will be called a *generator* of the optical plane.

It is now easy to show that if two distinct elements of a general line lie in an optical plane, then every element of the general line lies in the optical plane.

However, only optical and separation lines lie in an optical plane, and in this important respect it differs from an acceleration plane which contains also inertia lines.

We can easily prove that if e be a general line in an optica plane, and A be any element of the optical plane which does not lie in e, then there is one single general line through A in the optical plane which does not intersect e.

This is the parallel to e through A and the result given corresponds to the axiom of parallels in Euclidean geometry.

We can also easily show that Peano's axioms of order in a plane hold for an optical plane making use of the already defined meaning of *linearly between*.

We can also show that if A, B and C be three elements in an optical plane which do not all lie in one general line, and if D be the mean of A and B, then a general line through D parallel to BC intersects AC in an element which is the mean of A and C.

We can further prove a number of theorems in connexion with an optical plane analogous to those proved for an acceleration plane, and so bring our knowledge of the former up to a level with our knowledge of the latter.

It is easy to show that if three elements A_1, A_2, A_3 which do not lie in one general line lie in one optical plane, they determine the optical plane containing them and we are able to establish criteria by which we can say whether or no three elements do lie in an optical plane.

If we return to the consideration of our illustration there is, as already pointed out, a third type of plane besides those which represent acceleration and optical planes.

We have next to consider certain theorems in our geometry preparatory to the investigation of a type of plane analogous to a plane in our illustration which passes through the vertex of a cone, but has no other real point in common with the cone.

The first theorem is as follows:

If two optical parallelograms have a pair of opposite corners in common lying in an inertia line, then their separation diagonal lines are such that no element of the one is either *before* or *after* any element of the other.

If c and e be two such separation diagonal lines, then, as we have already seen, they have a common element, namely: the centre of the two optical parallelograms.

Further, any general line which intersects c and e in distinct elements must itself be a separation line.

It is easy to show that the separation lines c and e cannot both lie either in one acceleration plane or in one optical plane.

Before we can carry out the investigation of this third type of plane, we have to introduce a new postulate.

If A be any element, and a be an inertia line not containing A, while B is the element common to a and the α sub-set of A, then we shall speak of B as *the first element in a which is after A.*

Similarly, if C be the element common to a and the β sub-set of A, we shall speak of C as *the last element in a which is before A.*

The new postulate is as follows:

POSTULATE XVIII. **If a, b and c be three parallel inertia lines which do not all lie in one acceleration plane and A_1 be an element in a and if**

B_1 be the first element in b which is after A_1,

C_1 be the first element in c which is after A_1,

B_2 be the first element in b which is after C_1,

C_2 be the first element in c which is after B_1,

then the first element in a which is after B_2 and the first element in a which is after C_2 are identical.

From the physical standpoint this postulate looks to be among the simplest, and corresponds to the presumed optical fact that if we have three particles P, Q and R which are unaccelerated, at rest with respect to one another, and at the corners of a triangle, then if a flash of light starting from P goes directly round from P to Q to R and back again to P, and if another flash starting from P goes directly round from P to R to Q and back again to P, then if the two flashes start out simultaneously from P they return simultaneously to P.

The postulate is not, however, quite so simple as it appears, since it implies that we already know just what we mean when we say that the three particles are at rest with respect to one another and are unaccelerated.

We have, however, defined the meaning of inertia lines and their parallelism so that we have already overcome this difficulty.

If we take our ordinary illustration and represent the inertia lines by lines parallel to the axis of t, then we can see at once that the postulate holds for such lines, and it is not difficult to prove that it also holds if the lines representing the inertia lines are not parallel to the axis of t.

There is a (b) form of this postulate in which the word *last* is substituted for the word *first*, and the word *before* for the word *after*, but it is not independent and can readily be proved from the form given.

Definition. An inertia line and a separation line which are diagonal lines of the same optical parallelogram will be said to be *conjugate* to one another.

It is evident that if an inertia line and a separation line are conjugate they lie in one acceleration plane and intersect one another.

It is also evident that if A be an element lying in an inertia or separation line a in an acceleration plane P, then there is only one separation or inertia line through A and lying in P which is conjugate to a; since if two optical parallelograms lie in P and have a as a common diagonal line, then their other diagonal lines do not intersect (Post. XVI).

From this it also follows that if two intersecting separation lines b and c be both conjugate to the same inertia line a, then a, b and c cannot lie in the same acceleration plane, and we shall have a and b in one acceleration plane, say P, while a and c lie in another, say Q.

If O be the element of intersection of b and c, then O must lie both in P and Q and therefore in the inertia line a.

If A_1 be an element in a distinct from O there is one optical parallelogram in the acceleration plane P having O as centre and A_1 as one of its corners.

If A_2 be the corner opposite A_1, then there is an optical parallelogram in Q also having A_1 and A_2 as a pair of opposite corners and therefore having the same centre O.

The separation lines b and c will be the separation diagonal lines of the optical parallelograms in P and Q respectively, and so it follows that no element of b is either *before* or *after* any element of c.

By considerations similar to the above, we can see that if two intersecting inertia lines b and c be both conjugate to the same separation line a, then a and b must lie in one acceleration plane, while a and c lie in another distinct acceleration plane.

Further, if O be the element of intersection of b and c, then O lies in a.

In this case, however, since b and c are two intersecting inertia lines they must lie in one acceleration plane which must be distinct from both the others.

Again, it is clear that if a be an inertia or separation line lying in an acceleration plane P with a separation or inertia line b which is conjugate to a, then any general line c lying in P and parallel to b is also conjugate to a.

Also conversely, it is clear that if a be an inertia or separation line lying in an acceleration plane P with two distinct separation or inertia lines b and c which are each conjugate to a, then b and c must be parallel to one another.

By the help of Post. XVIII we are able to prove some other very important theorems regarding conjugacy such as the following:

If an inertia line c be conjugate to two intersecting separation lines d and e, then if A be any element of d and B be any distinct element of e, the general line AB is conjugate to a set of inertia lines which are parallel to c.

Also we can show that: if two inertia lines b and c intersect in an element A_1 and are both conjugate to a separation line a, then a is conjugate to every inertia line in the acceleration plane containing b and c which passes through the element A_1.

In the course of proving this result it is also shown that no element of a with the exception of A_1 is either *before* or *after* any element of either of the generators of the acceleration plane containing b and c which pass through A_1.

We can also prove the theorems:

If b and c be any two intersecting inertia lines, there is at least one separation line which is conjugate to both b and c.

Also if b and c be any two intersecting separation lines such that no element of the one is either *before* or *after* any element of the other, there is at least one inertia line which is conjugate to both b and c.

Further, if an inertia line a be conjugate to a separation line b, and if an inertia line a' be parallel to a while a separation line b' is parallel to b, and if a' and b' intersect one another, then a' is conjugate to b'.

Having proved these various theorems on conjugacy we are now able to go on with the proof of the existence of the third type of plane suggested by our illustration.

The first step is to prove the following theorem:

If a be a separation line, and B be any element which is not an element of a, and is neither *before* nor *after* any element of a, while c is a general line passing through B and parallel to a, then if A be any element of a, while C is an element of c distinct from B, a general line through C parallel to BA will intersect a.

We can easily show that no element of c is either *before* or *after* any element of a.

We can next show that: if A and B be two elements lying respectively in two parallel separation lines a and b which are such that no element of the one is either *before* or *after* any element of the other, and if A' be a second and distinct element in a, there is only one general line through A' and intersecting b which does not intersect the general line AB.

This is in fact the parallel to AB through A'.

We are ultimately able to prove that if a and b be two parallel separation lines such that no element of the one is either *before* or *after* any element of the other, and if c and d be any two non-parallel separation lines intersecting both a and b, then the aggregate consisting of all the elements in c and in all separation lines intersecting a and b which are parallel to c must be identical with the aggregate consisting of all the elements in d and in all separation lines intersecting a and b which are parallel to d.

Definition. If a and b be two parallel separation lines such that no element of the one is either *before* or *after* any element of the other, then the aggregate of all elements of all mutually parallel separation lines which intersect both a and b will be called a *separation plane.*

We can now show that if any two distinct elements of a general line lie in a separation plane, then every element of the general line lies in the separation plane.

It is easily seen that no element of a separation plane is either *before* or *after* any other element of it.

We can also prove various properties of a separation plane such as the analogues of Peano's axioms and the parallel axiom, and can bring our state of knowledge of a separation plane up to the level of our knowledge of acceleration and optical planes.

We can also obtain a criterion to show under what circumstances three elements which do not lie in one general line lie in a separation plane, and can prove that *any three elements which do not all lie in one general line must lie either in an acceleration plane, an optical plane, or a separation plane.*

We can also now introduce the term *general plane* as a common designation for all three types and may define it as follows:

Definition. If a and b be any two intersecting general lines, then the aggregate of all elements of the general line b, and of all general lines parallel to b which intersect a, will be called a *general plane.*

The axioms of Peano are now seen to hold for a general plane as does also the equivalent of the axiom of parallels.

It is not difficult to see that: if a and b be two intersecting general lines lying in a general plane P, and if through any element not lying in P two general lines a' and b' be taken respectively parallel to a and b, then if P' be the general plane determined by a' and b', the two general planes P and P' can have no element in common.

We can also show that if a' and b' intersect in O' there is a general line through O' and lying in P' which is parallel to any general line in P.

We have already given a definition of the parallelism of acceleration planes and are now in a position to give a definition of the parallelism of general planes which will include that of acceleration planes as a special case.

Definition. If P be a general plane, and if through any element A outside P two general lines be taken respectively parallel to two intersecting general lines in P, then the two general lines through A determine a general plane which will be said to be *parallel* to P.

We can readily see that: through any element outside a general plane P, there is one single general plane parallel to P.

We can also see that this general plane must be of the same kind as P.

We can further show that two distinct general planes which are parallel to a third general plane are parallel to one another.

If a pair of parallel general lines be both intersected by another pair of parallel general lines, then the four general lines will form a general parallelogram either in an acceleration plane, an optical plane, or a separation plane.

We can now prove that: if two general parallelograms have a pair of adjacent corners in common their remaining corners either lie in one general line or else form the corners of another general parallelogram.

This result is important when we come to develop the theory of congruence.

We have next to consider some theorems relating to the conjugacy of inertia and separation lines which lead on to the more general conception of the normality of general lines, of which indeed conjugacy is a special case.

We can show that if an inertia line a be conjugate to two separation lines intersecting in the element O and lying in a separation plane P, then a is also conjugate to every separation line lying in P and passing through O.

Further, if b be any such separation line, then a and b lie in an acceleration plane, say Q, and we can prove that there is one and only one separation line, say c, lying in P and passing through O which is conjugate to every inertia line in Q which passes through O.

Moreover, if R be the acceleration plane containing a and c, then we can also show that b is conjugate to every inertia line in R which passes through O.

It is thus seen that there is a reciprocal relation between the separation lines b and c.

We have now to introduce a new postulate.

All the postulates which have hitherto been introduced may be represented by ordinary geometric figures not involving more than three dimensions, but we have now to introduce one which we cannot thus represent and which therefore gives our geometry a sort of four-dimensional character. It is as follows:

POSTULATE XIX. **If P be any optical plane there is at least one element which is neither before nor after any element of P.**

If we consider our usual illustration, we have seen that an optical plane is represented by a tangent plane to one of our cones and any point in the three-dimensional space which contains these cones is either *before* or *after* some point of such a plane.

A point which actually lies in the plane will be *before* certain points of it and *after* certain others, but a point outside the plane cannot bear this double relation and must be either *before* or else *after* certain points of the plane.

We cannot represent in three dimensions the case of an element such as that whose existence is asserted in Post. XIX.

We may compare this with an analogous feature of an optical line in an acceleration plane.

In that case, any element in the acceleration plane is either *before* or *after* some element of the optical line, and if we take the special case of an element of the optical line itself, it is *before* certain elements of the optical line and *after* certain others.

The only way in which we can get an element which is neither *before* nor *after* any element of the optical line is if we go outside the acceleration plane.

It will be observed that the analogy is very close.

It is easy to show by the help of Post. XIX that: there are at least two distinct optical planes containing any optical line, and this might be taken as an alternative form of the postulate.

In proving a number of theorems from this stage on, our usual illustration fails us and we have to get along without it.

However, in many of the theorems which we require to consider, we can get a great part of the construction into three dimensions and the remaining portion is easy, but in any case the practice which we have had in proving theorems by means of the relations of *before* and *after* enables us to treat four-dimensional theorems without much extra difficulty.

The first theorem which we prove by means of this postulate is as follows:

If b be any separation line, and O be any element in it, there are at least two acceleration planes containing O, and such that b is conjugate to every inertia line in each of them which passes through O.

Another important result which we reach is that we may have an acceleration plane and a separation plane having only one element in common, and such that each inertia line through the common element in the former is conjugate to every separation line through it in the latter.

In ordinary three-dimensional geometry, of course, we cannot have two planes with only one point in common.

Again, the following theorem may be proved:

If two distinct acceleration planes P and P' have a separation line b in common, and if another separation line c intersecting b in the element O be conjugate to every inertia line in P which passes through O, then if c be conjugate to one inertia line in P' which passes through O, it is conjugate to every inertia line in P' which passes through O.

NORMALITY OF GENERAL LINES HAVING A COMMON ELEMENT

WE are now in a position to define what we mean when we say that a general line a is *normal* to a general line b which has an element in common with it.

Since a and b are not always general lines of the same kind, it is difficult to give any simple definition which will include all cases; but the introduction of the word "*normal*" is justified by the simplification which is thereby brought about in the statement of many theorems.

Moreover, once we have introduced coordinates, the condition of the normality of general lines is the same in all cases.

Only one case will be found to be strictly analogous to the normality of intersecting straight lines in ordinary geometry; namely, the case of two separation lines.

The other cases are so different from our ordinary ideas of lines "*at right angles*" that we do not propose to use this expression in connexion with them.

Thus for instance any optical line is to be regarded as being "normal to itself" and the use of the words "at right angles" would in this case clearly be an abuse of language.

The extension of the idea of normality from the cases of general lines having a common element to the cases of general lines which have not a common element is, however, quite analogous to the corresponding extension in ordinary geometry and will be made subsequently.

We are at present only concerned with the cases of general lines having a common element and shall naturally include among these that of an optical line being "normal" to itself.

Thus the complete definition of the normality of general lines having a common element is to be taken as consisting of the following four particular definitions A, B, C *and* D.

Definition A. An optical line will be said to be *normal to itself.*

Definition B. If an optical line a and a separation line b have an element O in common, and if no element of b with the exception of O be either *before* or *after* any element of a, then b will be said to be *normal* to a and a will be said to be *normal* to b.

Definition C. If an inertia line a and a separation line b be conjugate one to the other, then a will be said to be *normal* to b and b will be said to be *normal* to a.

Definition D. A separation line a having an element O in common with a separation line b will be said to be *normal* to b provided an acceleration plane P exists containing b, and such that every inertia line in P which passes through O is conjugate to a.

In this last case since there is one single inertia line in P which passes through O and is conjugate to b, it is evident that a and b lie in a separation plane.

Further, the result mentioned immediately prior to the introduction of Post. XIX shows that in this case the relation between a and b is a reciprocal one, so that b satisfies the definition of being normal to a provided a is normal to b.

It is clear from our definitions that for general lines which have an element in common we may have:

(i) A separation line normal to an inertia line, an optical line, or a separation line.

(ii) An optical line normal only to an optical line or a separation line.

(iii) An inertia line normal only to a separation line.

Again, if a be a separation line, and b a general line intersecting a and normal to it, then:

(i) If b be an inertia line, a and b lie in one acceleration plane.

(ii) If b be an optical line, a and b lie in one optical plane.

(iii) If b be a separation line, a and b lie in one separation plane.

If P be a separation plane, and if b be any separation line in P, and O be any element in b, we can easily show that there is one, and only one, separation line in P and passing through O which is normal to b.

Again, if instead P be an acceleration plane, and if b be any general line in P, and O be any element in b, then we know that if b be either an inertia or a separation line there is one and only one general line in P and passing through O which is conjugate and therefore normal to b.

Also from our definitions if b be an optical line there is still one and only one general line in P and passing through O which is normal to b: namely, b itself.

Thus we have the following general result:

If P be either a separation plane or an acceleration plane, and *if b be any general line in P and O be any element in b, then there is one and only one general line lying in P and passing through O which is normal to b.*

Now we have seen that if a separation line *a* be normal to a separation line *b* having an element in common with it, then *a* and *b* lie in a separation plane.

Thus two intersecting separation lines in an optical plane cannot be normal one to another.

Any separation line, however, which lies in an optical plane is normal to every optical line in the optical plane, since no element of the separation line except the element of intersection is either *before* or *after* any element of any optical line in the optical plane.

Since there is one, and only one, optical line which passes through any element of an optical plane and lies in the optical plane, we have the following result:

If P be an optical plane, and if b be any separation line in P and O be any element in b, then there is one, and only one, general line lying in P and passing through O which is normal to b.

If on the other hand b be an optical line lying in P, then every general line in P which passes through O (including b itself) is normal to b.

There is another very important theorem concerning the normality of general lines which is as follows:

If b and c be two distinct general lines having an element O in common, and if a general line a passing through O be normal to both b and c, then a is normal to every general line which passes through O and lies in the general plane containing b and c.

There are a number of special cases of this theorem which have already been mentioned, and some others which have not been referred to.

An enumeration of the different cases will be found in my larger work.

We can now prove that if *b* and *c* be two general lines intersecting in an element *O* and such that the one is normal to the other, and if they are respectively parallel to two other general lines *b′* and *c′* intersecting in an element *O′*, then of these latter two general lines the one is normal to the other.

We can now give the following definition:

Definition. A general line *b* will be said to be *normal* to a general line *c′* which has no element in common with it, provided that a general line b' taken through any element of c' parallel to b is normal to c' in the sense already defined.

Since any optical line is normal to itself, it follows from the above definition that any two parallel optical lines are to be regarded as normal to one another.

Definition. A general line a will be said to be *normal* to a general plane P provided a be normal to every general line in P.

It is evident that if a general line a be normal to two intersecting general lines in a general plane P, then a will be normal to P.

In case P be an optical plane it is clear that, according to the above definition, any generator of P is normal to P.

This is the only case in which a general line can be normal to a general plane which contains it.

In no other case can a general line which is normal to a general plane have more than one element in common with the latter.

It has already been pointed out that we may have an acceleration plane and a separation plane having only one element in common and such that each inertia line through the common element in the former is conjugate to every separation line through it in the latter.

It is evident now that we have here two general planes which are so related that any general line in the one is normal to any general line in the other.

In ordinary three-dimensional geometry two planes cannot be so related, and when we speak of one plane being normal to another, the normality is not of this complete character.

We shall therefore introduce the following definition:

Definition. If two general planes be so related that every general line in the one is normal to every general line in the other, the two general planes will be said to be *completely normal* to one another.

We are now able to prove that if P be any general plane and O be any element in it, there is at least one general plane passing through O and completely normal to P.

In fact if P be an acceleration plane the general plane completely normal to it whose existence we prove is a separation plane, if P be a separation plane the completely normal general plane is an acceleration plane, while if P be an optical plane, the completely normal general plane is an optical plane.

We can, however, go further and instead of taking O as any element in P we can prove a similar result if O be any element whatever.

Again, let O be any element and let S be any separation plane passing through O, while P is an acceleration plane passing through O and completely normal to S.

Let a be any separation line in S which passes through O, and let b be the one single separation line in S and passing through O which is normal to a.

Let c be any separation line passing through O and lying in P, and let d be the one single inertia line in P and passing through O which is normal to c.

Then both c and d are normal to both a and b and so *we have the three separation lines* a, b *and* c *all passing through* O *and each of them normal to the other two; while in addition to these we have the inertia line* d *also passing through* O *and normal to all three.*

This marks an important stage in the development of our theory as it suggests the possibility of setting up a system of coordinate axes one of which axes is of a different character from the remaining three.

Various other theorems concerning normality which are important in the logical development may now be proved and among these we mention the following:

If three general lines a, b and c have an element O in common, there is at least one general line passing through O which is normal to all three.

We are now in a position to take another important step in the development of our subject by defining *threefolds.*

In defining general lines and general planes, it was found necessary to define particular types separately, before any general definition could be given.

This difficulty does not occur in the case of general threefolds.

Definition. If a general line a and a general plane P intersect, then the aggregate of all elements of P and of all general planes parallel to P which intersect a will be called a *general threefold.*

It will be found that, just as there are three types of general line and three types of general plane, so there are three types of general threefold.

The distinction will be made hereafter, but there are many properties possessed in common by all three types and of which general proofs may be given.

Thus we can show that if two distinct elements of a general line lie in a general threefold, then every element of the general line lies in the general threefold.

Also if a general plane have three distinct elements in common with a general threefold, and if these three elements do not lie in one general line, then every element of the general plane lies in the general threefold.

Further, if a general line b lies in a general threefold W, and if A be any element lying in W but not in b, then the general line through A parallel to b also lies in W.

Also if a general plane P lies in a general threefold W, and if A be any element lying in W but not in P, then the general plane through A parallel to P also lies in W.

Moreover, we can show that if a general threefold W be determined by a general plane P and a general line a which intersects P, then if Q be any general plane lying in W and if b be any general line lying in W and intersecting Q, the general plane Q and the general line b also determine the same general threefold W.

It follows from this that any four distinct elements which do not all lie in one general plane determine a general threefold containing them; as do also any three distinct general lines having a common element and not all lying in one general plane.

We can also prove that if two distinct general planes P and Q lie in a general threefold W, then if P and Q have one element in common they have a second element in common and have therefore a general line in common.

Moreover, if P and Q have no element in common but both lie in W then they must be parallel.

Now we have already seen that we can have a separation plane S and an acceleration plane P having an element O in common and which are completely normal to one another.

We have also seen that in this case P and S cannot have a second element in common.

It follows that P and S cannot lie in one general threefold.

Now let a_1 and a_2 be any two distinct general lines lying in P and passing through O.

Then S and a_1 determine a general threefold, say W_1, while S and a_2 determine a general threefold, saw W_2.

Now W_1 and W_2 must be distinct, for if W_2 were identical with W_1, then W_1 would contain both a_1 and a_2 and would therefore contain P.

But W_1 contains S and so this is impossible.

Thus W_1 and W_2 are distinct general threefolds, each of which contains the separation plane S.

Since there are an infinite number of general lines lying in P and passing through O, it follows that there are an infinite number of general threefolds which all contain any separation plane S.

Similarly, there are an infinite number of general threefolds which all contain any acceleration plane P.

Without Post. XIX or some equivalent, we cannot from our remaining postulates show that there is more than one general threefold; for the proof of the existence of an acceleration plane which is completely normal to a separation plane depends upon Post. XIX.

We can now prove that: if a, b and c be any three distinct general lines having an element O in common but not all lying in one general plane, and if a general line d also passing through O be normal to a, b and c, then d is normal to every general line in the general threefold containing a, b and c.

Definition. A general line which is normal to every general line in a general threefold will be said to be *normal to the general threefold.*

We are now able to prove the existence of the three different types of general threefold and to give criteria by which we can tell in which kind of general threefold a given set of four elements (not all in one general plane) must lie.

The three different types may be defined as follows:

Definition. If a separation line a intersects a separation plane S and is normal to it, then the aggregate of all elements of S and of all separation planes parallel to S which intersect a will be called a *separation threefold.*

Definition. If an optical line a intersects a separation plane S and is normal to it, then the aggregate of all elements of S and of all separation planes parallel to S which intersect a will be called an *optical threefold.*

Definition. If an inertia line a intersects a separation plane S and is normal to it, then the aggregate of all elements of S and of all separation planes parallel to S which intersect a will be called a *rotation threefold.*

As regards the characteristic properties of these different types;

no element of a separation threefold is either *before* or *after* any other element of it, and so the only type of general lines which it contains are separation lines and the only type of general planes are separation planes.

As regards an optical threefold, through any element of it there is one single optical line lying in the optical threefold and all these optical lines are neutrally parallel to one another. These will be spoken of as *generators*.

An optical threefold contains separation lines and optical lines but no other type of general line, and it contains separation planes and optical planes but no other type of general plane.

As regards a rotation threefold, it contains all three types of general line and all three types of general plane.

Through any element of it there are an infinite number of optical lines, which we shall call *generators*, lying in the rotation threefold and in fact all the postulates which we introduced prior to Post. XIX hold for the set of elements contained in a rotation threefold without going outside that rotation threefold.

This is clearly not true for separation or optical threefolds.

We are now in a position to introduce a new postulate which limits the number of dimensions of our set of elements.

POSTULATE XX. **If W be any optical threefold, then any element of the set must be either before or after some element of W.**

This postulate should be compared with Post. XIX and the difference noted.

In Post. XIX the existence of an element neither *before* nor *after* any element of an optical plane is asserted, while in Post. XX the existence of an element neither *before* nor *after* any element of an optical threefold is denied.

As in the case of an optical line and an optical plane, so too in the case of an optical threefold, if any element be *after* one element of it and *before* another it must itself be an element of the optical threefold.

We are now able to prove that: if P be any general plane and O be any element of the set, there is one, and only one, general plane passing through O and completely normal to P.

We had previously shown that there was at least one such general plane and now we can show that it is unique.

We can also prove that:

(1) If *P* be any acceleration or separation plane and *O* be any element outside it, then the general plane through *O* and completely normal to *P* has one single element in common with *P*.

(2) If *P* be an optical plane and *O* be any element outside it, the optical plane through *O* and completely normal to *P* has an optical line in common with *P* if *O* be neither *before* nor *after* any element of *P*, but has no element in common with *P* if *O* be either *before* or *after* any element of *P*.

We can further prove that: if *W* be any general threefold and *O* be any element of the set, there is one, and only one, general line passing through *O* and normal to *W*.

Also if *a* be any general line and *O* be any element of the set there is one, and only one, general threefold passing through *O* and normal to *a*.

Further, we can show that: if *a* be a general line and *O* be any element in it while *W* is a general threefold passing through *O* and normal to *a*, then

(1) If *a* be an inertia line, *W* is a separation threefold.

(2) If *a* be a separation line, *W* is a rotation threefold.

(3) If *a* be an optical line, *W* is an optical threefold containing *a*.

We can now show that: if *W* be a general threefold and *A* be any element outside it, then any general line through *A* is either parallel to a general line in *W* or else has one single element in common with *W*.

Since a separation threefold contains only separation lines, while an optical threefold contains only separation lines and a system of parallel optical lines, it follows from the last result that: every inertia line and every optical line intersects every separation threefold, while every inertia line and every optical line which is not parallel to a generator of an optical threefold intersects the optical threefold.

Again, we can show that if *W* be a general threefold and *P* be a general plane which does not lie in *W*, then if *P* has one element in common with *W*, it has a general line in common with *W*.

Further, if W_1 and W_2 be two distinct general threefolds having an element *A* in common, then they have a general plane in common.

We can also prove that: if *W* be a general threefold and *O* be any element outside it, and if through *O* there pass three general lines *a*, *b* and *c* which do not all lie in one general plane and which are respectively parallel to three general lines in W, then a, b and c determine a general threefold W', such that every general line in W' is parallel to a general line in W.

We accordingly introduce the following definition:

Definition. If W be a general threefold, and if through any element A outside W three general lines be taken not all lying in one general plane but respectively parallel to three general lines in W, then the general lines through A determine a general threefold which will be said to be *parallel* to W.

It is clear that a general threefold can only be parallel to another general threefold if they are both of the same type.

It is also easy to see that if A be any element outside a general threefold W, there is only one general threefold passing through A and parallel to W.

Further, if two distinct general threefolds are both parallel to the same general threefold they are parallel to one another.

There are various cases of partial normality of general planes to general planes and general threefolds, and of general threefolds to general threefolds considered in my larger work, but which we do not propose to go into here.

There is one very important theorem we shall now enunciate which is required in the treatment of the theory of congruence: the subject we propose next to take up. It is as follows:

If A, B, C, D be the corners of an optical parallelogram (AC being the inertia diagonal line), and if A, B', C, D' be the corners of a second optical parallelogram, while A', B', C', D' are the corners of a third optical parallelogram whose diagonal line $A'C'$ is conjugate to BD, then A', B, C', D will be the corners of a fourth optical parallelogram.

The importance of this theorem lies in the fact that it enables us to show the unique character of the relations of certain elements in respect of congruence.

THEORY OF CONGRUENCE

We are now in a position to consider the problems of *congruence* and *measurement* in our system of geometry.

The first point to be considered is the *congruence* of pairs of elements and we shall find that there are several cases which have to be considered separately.

Two distinct elements A and B will be spoken of briefly as a *pair* and will be denoted by the symbols (A, B) or (B, A).

The order in which the letters are written will be taken advantage of in order to symbolize a certain correspondence between the elements of pairs, as we shall shortly explain.

Since any two distinct elements determine a general line containing them, there will always be one general line associated with any given pair, but different pairs will be associated with the same general line.

If we set up a correspondence between the elements of a pair (A, B) and a pair (C, D) we might either take C to correspond to A and D to B, or else take D to correspond to A and C to B.

The first of these might be symbolized briefly by:

$$(A, B) \text{ corresponds to } (C, D),$$

or

$$(B, A) \text{ corresponds to } (D, C).$$

The second might be symbolized by:

$$(A, B) \text{ corresponds to } (D, C),$$

or

$$(B, A) \text{ corresponds to } (C, D).$$

If we consider the case of pairs which have a common element, say (A, B) and (A, C), and if

$$(A, B) \text{ corresponds to } (A, C),$$

then the element A corresponds to itself and will be said to be *latent*.

Now the congruence of pairs is a correspondence which can be set up in a certain way between certain pairs lying in general lines of the same type.

The correspondence is set up in virtue of certain similarities of relationship.

In dealing with this subject, it will be found convenient to have a systematic notation for optical parallelograms, so that we may be able to distinguish how the different corners are related.

If A, B, C, D be the corners of an optical parallelogram we shall use the notation $A\overline{BC}D$ when we wish to signify that the corners A and D lie in the inertia diagonal line and that A is *before* D, while B and C lie in the separation diagonal line so that the one is neither *before* nor *after* the other.

If O be the centre of the optical parallelogram $A\overline{BC}D$, it is obvious that O will be *after* A and *before* D.

A pair (A, B) will be spoken of as an *optical pair*, an *inertia pair* or a *separation pair* according as AB is an optical, an inertia, or a separation line.

We shall first give a definition of the congruence of inertia pairs having a latent element.

Definition. If $A_1\overline{BC}D_1$ and $A_2\overline{BC}D_2$ be optical parallelograms having the common pair of opposite corners B and C and the common centre O, then the inertia pair (O, D_1) will be said to be *congruent* to the inertia pair (O, D_2).

This will be written

$$(O, D_1)\ (\equiv)\ (O, D_2).$$

Similarly the inertia pair (O, A_1) will be said to be *congruent* to the inertia pair (O, A_2).

If (O, D_1) be any inertia pair and a be any inertia line intersecting OD_1 in O, then the above definition enables us to show that there is one and only one element, say X, in a which is distinct from O and such that:

$$(O, D_1)\ (\equiv)\ (O, X).$$

The unique character of the element X is proved by means of the theorem stated at the end of the last section.

Again, if (O, D_1), (O, D_2) and (O, D_3) be inertia pairs such that:

$$(O, D_1)\ (\equiv)\ (O, D_2),$$

and

$$(O, D_2)\ (\equiv)\ (O, D_3),$$

it is easy to show that:

$$(O, D_1)\ (\equiv)\ (O, D_3).$$

In order to see this we have only to remember that whether the inertia lines OD_1, OD_2, OD_3 all lie in one acceleration plane or in one rotation threefold, there must be at least one separation line passing through O and normal to all three.

Thus *for inertia pairs having a latent element the relation of congruence is a transitive relation.*

We shall next consider the congruence of separation pairs having a latent element.

This case differs somewhat from the one we have considered.

While two intersecting inertia lines always lie in an acceleration plane, two intersecting separation lines may lie either in a separation plane, an optical plane, or an acceleration plane.

An inertia line can only be conjugate to two intersecting separation lines if these lie in a separation plane, and so if we were to give a definition of the congruence of separation pairs having a latent element which was strictly analogous to that given for inertia pairs, such a definition would be incomplete.

It is, however, possible to give a definition based on that already given for inertia pairs which will include all cases.

In order to avoid complication we shall first explain what we mean by an inertia pair being "conjugate" to a separation pair or a separation pair being "conjugate" to an inertia pair.

Definition. If $A\overline{BC}D$ be an optical parallelogram and O be its centre, then the inertia pairs (O, D) and (O, A) will be spoken of as *conjugates* to the separation pairs (O, B) and (O, C) and also conversely.

The pair (O, D) will be called an *after-conjugate* to the pairs (O, B), (O, C), while (O, A) will be called a *before-conjugate* to the pairs (O, B), (O, C).

Further, either of the separation pairs (O, B), (O, C) will be called an *after-conjugate* to (O, A) and a *before-conjugate* to (O, D).

Now we know that there are an infinite number of acceleration planes which contain any given separation line and so there are always inertia pairs which are conjugate to any given separation pair.

Knowing this we can give the following definition of the "congruence" of separation pairs having a latent element.

Definition. If (O, B_1) and (O, B_2) be separation pairs, and if (O, D_1) and (O, D_2) be inertia pairs which are after-conjugates to (O, B_1) and (O, B_2) respectively, then if $(O, D_1)\ (\equiv)\ (O, D_2)$ we shall say that (O, B_1) is congruent to (O, B_2) and shall write this:

$$(O, B_1)\ \{\equiv\}\ (O, B_2).$$

It is easy to see that the congruence of (O, B_1) to (O, B_2) is independent of the particular after-conjugates to (O, B_1) and (O, B_2) which we may select.

From the corresponding result for the case of inertia pairs we can prove directly that if (O, B_1), (O, B_2) and (O, B_3) be separation pairs such that:

$$(O, B_1)\{\equiv\}(O, B_2),$$

and

$$(O, B_2)\{\equiv\}(O, B_3),$$

then

$$(O, B_1)\{\equiv\}(O, B_3),$$

or *for separation pairs having a latent element, the relation of congruence is a transitive relation.*

Again, if (O, B) be any separation pair and a be any separation line passing through O, there are two and only two elements, say X_1 and Y_1, in a which are distinct from O and such that:

$$(O, B)\{\equiv\}(O, X_1),$$

and

$$(O, B)\{\equiv\}(O, Y_1).$$

In fact if $A\overline{BC}D$ be an optical parallelogram and O be its centre we observe that according to our definitions we have:

$$(O, B)\{\equiv\}(O, C),$$

but not

$$(O, A)(\equiv)(O, D).$$

The reason why we make this distinction is that in the separation pairs we have O neither *before* nor *after* B and also O neither *before* nor *after* C, while in the inertia pairs we have O *after* A and O *before* D.

Thus in the first case the relations are alike in respect of *before* and *after*, while in the second case the relations are different.

The question now arises as to the "congruence" of optical pairs.

In this case constructions such as those by which we defined the congruence of inertia and separation pairs having a latent element entirely fail and there is nothing at all analogous to them.

We are thus led to regard optical pairs as not determinately comparable with one another in respect of congruence, except when they lie in the same, or in parallel optical lines.

In fact to suppose otherwise would be to destroy the symmetry of our geometry, and it is on this account that when we make use of our usual illustration of conical order by means of cones, we get distortion; since the generators of the cones are ordinary straight lines, and a portion of any one is comparable with a portion of any other in respect of length.

As regards the "congruence" of pairs lying in the same general line, no definition has yet been given except for the very special case of inertia or separation pairs having a latent element; while no definition whatever has been given of the "congruence" of pairs lying in parallel general lines.

The method by which we do this is by employing the properties of general parallelograms.

Definition. A pair (A, B) will be said to be *opposite* to a pair (C, D), if, and only if, the elements A, B, C, D form the corners of a general parallelogram in such a way that AB and CD are one pair of opposite sides, while AC and BD are the other pair of opposite sides.

This will be denoted by the symbols:

$$(A, B) \square (C, D).$$

It will be observed that the use of the symbol $\square$ implies that the pairs (A, B) and (C, D) lie in distinct general lines which are parallel to one another.

If, however, we have

$$(A, B) \square (C, D),$$

and

$$(E, F) \square (C, D),$$

then the pairs (A, B) and (E, F) may lie either in the same or in parallel general lines.

If (A, B) and (E, F) do not lie in the same general line, then it follows from a theorem already mentioned (p. 43) that we may write:

$$(A, B) \square (E, F).$$

We can now prove the following theorem:

If (A, B), (A', B') and (C, D) be pairs such that:

$$(A, B) \square (C, D),$$

and

$$(A', B') \square (C, D),$$

and if (C', D') be any other pair such that:

$$(A, B) \square (C', D'),$$

and which does not lie in the general line $A'B'$, then we shall also have:

$$(A', B') \square (C', D').$$

We can now introduce the following definition:

Definition. A pair (A, B) will be said to be *co-directionally congruent* to a pair (A', B') provided a pair (C, D) exists such that:

$$(A, B) \square (C, D),$$

and

$$(A', B') \square (C, D).$$

The theorem above enunciated shows that we are at liberty to replace the pair (C, D) by any other pair (C', D') such that:

$$(A, B) \sqsubset (C', D'),$$

provided that (C', D') does not lie in the general line $A'B'$.

We shall symbolize the co-directional congruence of (A, B) to (A', B') thus:

$$(A, B) \mid\equiv\mid (A', B').$$

It will be seen that $(A, B) \sqsubset (A', B')$ implies that

$$(A, B) \mid\equiv\mid (A', B')$$

but that the latter does not imply the former except when AB and $A'B'$ are distinct general lines.

We can easily prove that if

$$(A, B) \mid\equiv\mid (C, D),$$

and

$$(C, D) \mid\equiv\mid (E, F),$$

then

$$(A, B) \mid\equiv\mid (E, F),$$

or *the relation of co-directional congruence of pairs is a transitive relation.*

It is now possible to give a general definition of the congruence of inertia or separation pairs.

This is done by combining co-directional congruence with congruence in which an element is latent.

This may be compared to combining translation with rotation, but differs in this, that it does not imply the motion of any rigid body.

Definition. An inertia pair (A_1, B_1) will be said to be *congruent* to an inertia pair (A_2, B_2) provided an inertia pair (A_2, C_2) exists such that:

$$(A_1, B_1) \mid\equiv\mid (A_2, C_2),$$

and

$$(A_2, B_2) (\equiv) (A_2, C_2).$$

Definition. A separation pair (A_1, B_1) will be said to be *congruent* to a separation pair (A_2, B_2) provided a separation pair (A_2, C_2) exists such that:

$$(A_1, B_1) \mid\equiv\mid (A_2, C_2),$$

and

$$(A_2, B_2) \{\equiv\} (A_2, C_2).$$

We shall denote the generalized congruence of inertia or of separation pairs by the symbol $\equiv$ thus:

$$(A_1, B_1) \equiv (A_2, B_2).$$

We shall also use the same symbol to denote the congruence of optical pairs, but in the latter case it is to be regarded as simply equivalent to the symbol $|\equiv|$ since the only congruence of optical pairs is taken to be co-directional.

In my larger work several theorems are proved prior to the introduction of the general definitions of the congruence of inertia and of separation pairs.

These theorems enable us to show that this general congruence is of a reciprocal character so that

$$(A_1, B_1) \equiv (A_2, B_2),$$

implies that

$$(A_2, B_2) \equiv (A_1, B_1),$$

both for inertia and for separation pairs and also

$$(A_1, B_1) \equiv (A_2, B_2),$$

implies that

$$(B_1, A_1) \equiv (B_2, A_2),$$

for both types of pairs.

Further, both for inertia and for separation pairs it is easy to see that

$$(A, B) \equiv (A, B),$$

and we can also prove that if

$$(A_1, B_1) \equiv (A_2, B_2),$$

and

$$(A_2, B_2) \equiv (A_3, B_3),$$

then

$$(A_1, B_1) \equiv (A_3, B_3).$$

Thus for inertia or separation pairs the general relation of congruence is a transitive relation.

Again, if (A, B) be a separation pair we can show that

$$(A, B) \equiv (B, A).$$

We have not, however, a corresponding result in the case of either inertia or optical pairs since the elements in such pairs are asymmetrically related.

We can easily prove that if (A_1, B_1), (A_2, B_2), (B_1, C_1), (B_2, C_2) be pairs such that

$$(A_1, B_1) \,|\equiv|\, (A_2, B_2),$$

and

$$(B_1, C_1) \,|\equiv|\, (B_2, C_2),$$

then if C_1 be distinct from A_1 we shall also have

$$(A_1, C_1) \,|\equiv|\, (A_2, C_2).$$

We can also show that if (A_1, B_1), (A_2, B_2), (B_1, C_1), (B_2, C_2) be inertia or separation pairs such that

$$(A_1, B_1) \equiv (A_2, B_2),$$

and

$$(B_1, C_1) \equiv (B_2, C_2),$$

then if B_1 be linearly between A_1 and C_1 while B_2 is linearly between A_2 and C_2 we shall also have

$$(A_1, C_1) \equiv (A_2, C_2).$$

A similar result also holds for optical pairs, but in that case the congruence must of course be co-directional.

Again, we can show that if A and B be two distinct elements and E be any element in AB distinct from A and B, while F is an element in AB such that

$$(A, E) \mid\equiv\mid (F, B),$$

then we shall have $\qquad (A, F) \mid\equiv\mid (E, B).$

Definitions. If A and B be two distinct elements, then the set of all elements lying linearly between A and B will be called the *segment AB*.

The elements A and B will be called the *ends of the segment*, but are not included in it.

The set of elements obtained by including the ends will be called a *linear interval.*

If A and B be two distinct elements, then the set of elements such as X where B is linearly between A and X may be called the *prolongation of the segment AB beyond B.*

Such a set of elements may also be spoken of as a *general half-line.*

The element B will be called the *end* of the general half-line.

We shall describe segments and general half-lines as *optical*, *inertia*, or *separation*, according as they lie in optical, inertia, or separation lines.

If A, B, C be three distinct elements which do not all lie in one general line, then the three segments AB, BC, CA, together with the three elements A, B, C, will be called a *general triangle* or briefly a *triangle* in an acceleration, optical, or separation plane, as the case may be.

The elements A, B, C will be called the *corners*, while the segments AB, BC, CA will be called the *sides* of the general triangle.

We can now prove the following very important theorem:

If A_1, B_1, C_1 be the corners of a triangle in a separation plane P_1 and A_2, B_2, C_2 be the corners of a triangle in a separation plane P_2, and if further $\qquad (C_1, A_1) \equiv (C_2, A_2),$

$$(C_1, B_1) \equiv (C_2, B_2),$$

while B_1C_1 is normal to A_1C_1, and B_2C_2 is normal to A_2C_2, then we shall also have $$(A_1, B_1) \equiv (A_2, B_2).$$

It will be observed that this theorem is analogous to the fourth proposition of Euclid for the special case of right-angled triangles and we are able to prove it without superposition.

We can also prove two converses of this theorem as follows:

If A_1, B_1, C_1 be the corners of a triangle in a separation plane P_1, and A_2, B_2, C_2 be the corners of a triangle in a separation plane P_2, then:

(1) If $$(C_1, A_1) \equiv (C_2, A_2),$$ $$(A_1, B_1) \equiv (A_2, B_2),$$

while B_1C_1 is normal to A_1C_1 and B_2C_2 is normal to A_2C_2, we shall also have $$(C_1, B_1) \equiv (C_2, B_2).$$

(2) If $$(A_1, B_1) \equiv (A_2, B_2),$$ $$(A_1, C_1) \equiv (A_2, C_2),$$ $$(B_1, C_1) \equiv (B_2, C_2),$$

while A_1C_1 is normal to B_1C_1, then we shall also have A_2C_2 normal to B_2C_2.

A further theorem may now be proved whose importance consists in this: that it is equivalent to one of the assumptions used by Professor Veblen in his treatment of the subject of congruence in ordinary Euclidean geometry.

It is as follows:

If A_1, B_1, C_1 be the corners of a triangle in a separation plane P_1, and A_2, B_2, C_2 be the corners of a triangle in a separation plane P_2, and if D_1 be an element in B_1C_1 such that C_1 is linearly between B_1 and D_1, while D_2 is an element in B_2C_2 such that C_2 is linearly between B_2 and D_2; and if further

$$(A_1, B_1) \equiv (A_2, B_2),$$ $$(B_1, C_1) \equiv (B_2, C_2),$$ $$(C_1, A_1) \equiv (C_2, A_2),$$ $$(B_1, D_1) \equiv (B_2, D_2),$$

we shall also have $$(A_1, D_1) \equiv (A_2, D_2).$$

Definitions. If O and X_0 be two distinct elements in a separation plane S, then the set of all elements in S such as X, where $$(O, X) \equiv (O, X_0),$$

will be called a *separation circle.*

The element O will be called the *centre* of the separation circle.

Any one of the linear intervals such as OX will be called a *radius* of the separation circle.

If X_1 and X_2 be two elements of the separation circle such that X_1X_2 passes through O, then the linear interval X_1X_2 will be called a *diameter* of the separation circle.

Any element which lies in a radius but which is not an element of the separation circle itself will be said to lie *inside* or in the *interior* of the separation circle.

Any element which lies in S but not in a radius will be said to lie *outside* or *exterior to* the separation circle.

It is easy to see that any general line b in a general plane P divides the remaining elements of P into two sets such that if A and C be any two elements of opposite sets, then b will intersect AC in an element linearly between A and C; while if A and A' be two elements of the same set, then b will not intersect AA' in any element linearly between A and A'.

If elements X and Y lie in the general plane P but not in the general line b, they will be said to lie *on the same side* of b if they both lie in the same set and will be said to lie *on opposite sides* of b if X lies in one of the sets and Y in the other set.

We can now prove that if a separation circle and a separation line both lie in a separation plane S, they cannot have more than two elements in common.

We can also prove that if a separation circle in a separation plane S pass through an element A which is inside and another element B which is outside a second separation circle in S, then the two separation circles have two elements in common which lie on opposite sides of the separation line AB.

This theorem also corresponds to one of Veblen's fundamental assumptions.

We can also prove a theorem which is the equivalent of the well-known *Axiom of Archimedes*. It is as follows:

If A_0, A_1 and C be three distinct elements such that A_1 is linearly between A_0 and C, and if A_2, A_3, A_4 ... be elements such that A_1 is linearly between A_0 and A_2,

A_2 is linearly between A_1 and A_3,

....

....

and such that $(A_0, A_1) \equiv (A_1, A_2) \equiv (A_2, A_3) \dots,$

then there are not more than a finite number of the elements $A_1, A_2, A_3 \ldots$ linearly between A_0 and C.

We shall now give the final postulate of our system which is equivalent to the axiom of Dedekind, and which therefore renders the set of elements a continuum in the mathematical sense.

POSTULATE XXI. **If all the elements of an optical line be divided into two sets such that every element of the first set is before every element of the second set, then there is one single element of the optical line which is not before any element of the first set and is not after any element of the second set.**

Since an element is neither *before* nor *after* itself it is evident that this one single element may belong either to the first or second set.

It is possible to prove that the Dedekind property holds also for inertia and for separation lines, but in the case of the latter it is necessary to formulate it somewhat differently, since no element of a separation line is either *before* or *after* any other element.

The necessary modification is given in my larger work.

Definitions. If (A, B) and (C, D) be inertia or optical pairs in which B is *after* A and D *after* C, or if (A, B) and (C, D) be separation pairs, then:

(1) If $(A, B) \equiv (C, D)$ we shall say that the segment AB is *equal to* the segment CD.

(2) If $(A, B) \equiv (C, E)$ where E is an element linearly between C and D we shall say that the segment AB is *less than* the segment CD.

(3) If $(A, B) \equiv (C, F)$ where F is any element such that D is linearly between C and F we shall say that the segment AB is *greater than* the segment CD.

In the case of separation or inertia segments we must always have either AB is equal to CD,

or AB is less than CD,

or AB is greater than CD.

In the case of optical segments, however, this is only true provided they lie in the same or parallel optical lines.

Again, if (A, B) and (C, D) be inertia or optical pairs in which B is *after* A and D *after* C, or if they be separation pairs, and if

E, F, G be elements such that F is linearly between E and G while

$$(A, B) \equiv (E, F),$$

and

$$(C, D) \equiv (F, G),$$

we shall say that *the length of the segment EG is equal to the sum of the lengths of the segments AB and CD.*

It is evident that the lengths of two optical segments can only have a sum in this sense provided they lie in the same or parallel optical lines, whereas the lengths of two inertia segments or two separation segments always have a sum.

Having thus introduced the idea of the length of a segment being equal to the sum of the lengths of two others, we can obviously have any *multiple* and also (as follows from the remarks on p. 36) any *sub-multiple* of a given segment: using the terms "multiple" and "sub-multiple" in the ordinary sense.

We can also clearly have a segment equal to any proper or improper fractional part of the given segment, and, by making use of our equivalents of the Archimedes and Dedekind axioms along with the corresponding properties of *real numbers*, we can complete the whole theory of representing lengths along a general line by means of real numbers.

The logical details of this will be found, for instance, in Pierpont's *Theory of Functions of Real Variables*, vol. I, chapters I and II.

The criterion of proportion given by Euclid is clearly applicable in our geometry, and we can readily show that separation segments are proportional to their conjugate inertia segments, etc.

We have now reached a stage in our investigation where we are able to show that: *the geometry of a separation threefold is formally identical with the ordinary (Euclidean) geometry of three dimensions.*

The way in which this is done in my larger work is by showing that, in a separation threefold, a set of propositions holds which had already been shown by Veblen to be sufficient as a basis for the ordinary three-dimensional (Euclidean) geometry.

From this stage on, we are evidently at liberty to make use of any known theorem of ordinary geometry and apply it in the geometry of a separation threefold.

Thus if A, B, C be the corners of a triangle in a separation plane, such that BC is normal to AC, then the theorem of Pythagoras shows that

$$(AB)^2 = (BC)^2 + (AC)^2.$$

For the details as to how this development is carried out, we must refer the reader to the works of Veblen and others.

We have next to consider some congruence theorems in optical and acceleration planes, which in some respects are quite different from the corresponding theorems in separation planes.

The following theorem is one which shows up this difference in a very striking way:

If O and X_0 be two distinct elements in a separation line lying in an optical plane P, then the set of elements in P such as X where OX is a separation line and

$$(O, X) \equiv (O, X_0)$$

consists of a pair of parallel optical lines.

In other words a pair of parallel optical lines is the analogue in an optical plane of a circle: that is to say, in so far as an analogue exists.

It is obvious that two such loci in the same optical plane can never intersect although they may have an optical line in common.

Further, instead of a single centre, as in the case of a circle, we have an optical line such that any element in it may be regarded as a centre.

If A, B and C be the corners of a triangle in an optical plane P such that AB, BC and CA are all separation lines and we take optical lines in P passing through A, B and C, then these optical lines will intersect BC, CA and AB respectively in elements which we shall denote by A', B' and C' respectively.

It is easy to prove that, in all cases, one and only one of the three elements A', B', C' lies linearly between a pair of the corners A, B, C.

Now let us consider the case, for instance, where A' is linearly between B and C.

It follows at once from the above-mentioned theorem that

$$(B, A) \equiv (B, A'),$$

and

$$(C, A) \equiv (C, A'),$$

a result which may be expressed in the following form:

If all three sides of a triangle in an optical plane be separation segments, then the sum of the lengths of a certain two of the sides is equal to that of the third side.

The geometry of an optical plane, although it shows some very remarkable features, is yet, in many respects, simpler than that of either a separation plane or an acceleration plane.

It is next necessary to consider some congruence properties of triangles in acceleration planes.

If A_1, B_1, C_1 be the corners of a triangle in an acceleration plane P_1, and A_2, B_2, C_2 be the corners of a triangle in an acceleration plane P_2, and if further, B_1C_1 be a separation line which is normal to the inertia line A_1C_1, while B_2C_2 is a separation line which is normal to the inertia line A_2C_2, then:

(1) If $(C_1, A_1) \equiv (C_2, A_2)$,

and $(C_1, B_1) \equiv (C_2, B_2)$,

we shall either have $(A_1, B_1) \equiv (A_2, B_2)$,

or else both A_1B_1 and A_2B_2 will be optical lines.

(2) If $(A_1, C_1) \equiv (C_2, A_2)$,

and $(C_1, B_1) \equiv (C_2, B_2)$,

we shall either have $(A_1, B_1) \equiv (B_2, A_2)$,

or else both A_1B_1 and B_2A_2 will be optical lines.

It will be observed that this theorem corresponds to the fourth proposition of Euclid for the special case where the given sides of each of the two triangles are normal to one another.

It will be noticed that when A_1B_1 and A_2B_2 are optical lines, we are not at liberty to assert the congruence of the corresponding pairs, although when they are not optical lines we can assert their congruence.

It may even happen, if A_1B_1 and A_2B_2 are optical lines, that the segment A_1B_1 is a part of the segment A_2B_2.

It will be seen that this theorem is considerably more complicated than the corresponding theorem for the case of separation planes, and the same will be found to hold for other congruence theorems in acceleration planes.

If we express these theorems in terms of segments instead of pairs, there is not so much complication, but by doing so we, to some extent, lose sight of certain *before* and *after* relations.

It is nevertheless frequently desirable to express these theorems in terms of segments instead of pairs, but even then, optical segments are exceptional.

The method by which the above theorem is proved, is by showing that these triangles in acceleration planes may each be correlated with a triangle in a separation plane, and the known congruence properties of these latter triangles are then used to establish the congruence properties of the former.

If we consider the triangle whose corners are A_1, B_1, C_1, we have B_1C_1 a separation line, and A_1C_1 is an inertia line normal to B_1C_1, while A_1B_1 may be: (i) a separation line, (ii) an inertia line, (iii) an optical line.

In case (i) we obtain a triangle whose corners are B_1, C_1 and E_1 lying in a separation plane, and such that E_1B_1 is normal to E_1C_1 and in which accordingly we must have the segment relation

$$(B_1C_1)^2 = (E_1B_1)^2 + (E_1C_1)^2.$$

This triangle is related to the one whose corners are A_1, B_1, C_1 in such a way that

$$(E_1, B_1) \equiv (B_1, A_1),$$

while (C_1, E_1) is a before- or after-conjugate to (C_1, A_1).

Thus taking segments instead of pairs we get

$$(B_1C_1)^2 = (B_1A_1)^2 + (\text{conjugate } C_1A_1)^2,$$

which we may write in the form

$$(B_1A_1)^2 = (B_1C_1)^2 - (\text{conjugate } C_1A_1)^2 \quad \text{.........(1).}$$

Again, if we consider case (ii), we obtain a triangle whose corners are C_1, B_1, F_1 lying in a separation plane and such that B_1F_1 is normal to B_1C_1.

Thus we must have the segment relation

$$(C_1F_1)^2 = (B_1C_1)^2 + (B_1F_1)^2.$$

This triangle is related to the one whose corners are A_1, B_1, C_1 in such a way that (C_1, F_1) is a before- or after-conjugate to (C_1, A_1), while (B_1, F_1) is a before- or after-conjugate to (B_1, A_1), and so taking segments instead of pairs, we get

$$(\text{conjugate } C_1A_1)^2 = (B_1C_1)^2 + (\text{conjugate } B_1A_1)^2.$$

This may be written in the form

$$-(\text{conjugate } B_1A_1)^2 = (B_1C_1)^2 - (\text{conjugate } C_1A_1)^2 \quad \text{...(2).}$$

In case (iii) we obviously have

$$0 = (B_1C_1)^2 - (\text{conjugate } C_1A_1)^2 \quad \text{...........(3).}$$

We now see that (1), (2) and (3) constitute the complete analogue of the theorem of Pythagoras in an acceleration plane.

If now we consider a triangle whose corners are A_1, B_1, C_1 and which lies in an optical plane, then if B_1C_1 be a separation line, and A_1C_1 be normal to B_1C_1, we know that A_1C_1 must be an optical line, while A_1B_1 must be another separation line.

Now we have seen that

$$(B_1, A_1) \equiv (B_1, C_1),$$

and so taking segments instead of pairs, we see that

$$(B_1A_1)^2 = (B_1C_1)^2 \ldots\ldots\ldots\ldots\ldots\ldots\ldots(4).$$

This is the analogue of the theorem of Pythagoras in an optical plane.

Considering now equations (1), (2), (3) and (4), we observe that *the modifications which take place in the theorem of Pythagoras are such that when any side of the triangle becomes an inertia segment, the corresponding square is replaced by the negative square of the conjugate of this inertia segment, while if any side becomes an optical segment, the corresponding square is replaced by zero.*

It is easy to prove the converse of this generalized Pythagorean theorem and to show that when the sides of the triangle are of the specified kinds and the specified relations hold, then B_1C_1 is normal to A_1C_1.

We can now show that: if A, B, C be the corners of a general triangle all of whose sides are segments of one kind, then:

(1) *If the triangle lies in a separation plane, the sum of the lengths of any two sides is greater than that of the third side.*

(2) *If the triangle lies in an optical plane, the sum of the lengths of a certain two sides is equal to that of the third side.*

(3) *If the triangle lies in an acceleration plane, the sum of the lengths of a certain two sides is less than that of the third side.*

These remarkable results show how necessary it is in studying this subject to lay down precisely what we take as our fundamental postulates, for it is clear that it would have been incorrect to have defined a linear segment in this geometry as: "the shortest distance between its extremities."

INTRODUCTION OF COORDINATES

If we take any element O of the set as origin, we have already seen that we may obtain systems of four general lines through O, say OX, OY, OZ, OT, which are mutually normal to one another.

Three of these, say OX, OY, OZ, will be separation lines, while the fourth, OT, will be an inertia line.

The three separation lines OX, OY, OZ will determine a separation threefold, say W, and OT will be normal to it.

If we select any arbitrary separation segment as a unit of length and associate the number zero with the element O, we may associate every other element of OX, OY, OZ with a real number, positive or negative, corresponding to the length of the segment of which that element is one end and the origin O is the other.

In this way we set up a coordinate system in W which will be quite similar to that with which we are familiar.

Since all the theorems of ordinary Euclidean geometry hold for a separation threefold, the length of a segment in W will be given by the ordinary Cartesian formula.

Again, not confining our attention merely to the elements of W, let A be any element of the whole set.

Then A must either lie in OT, or else there is an inertia line through A parallel to OT, and as we have already seen (p. 54), this inertia line will intersect W in some element, say N.

Further, AN must be normal to W.

Now if A does not lie in W, there will be a separation threefold, say W', passing through A and parallel to W, and the inertia line OT must intersect W' in some element, say M.

Further, since W' is parallel to W, both OT and AN must be normal to W'.

Thus if OM and NA are distinct, MA and ON must both be separation lines normal to OM, and so, since OM and NA lie in an acceleration plane, we must have MA parallel to ON.

Now we may select a unit inertia segment, just as we selected a unit separation segment, and with each element of OT distinct from O, we may associate a real number positive or negative, corresponding to the length of the segment of which that element is one end and the origin O is the other.

We shall suppose this correspondence to be set up in such a way that a positive real number corresponds to any element which is after O, and a negative real number to any element which is before O.

As regards the relationship between the unit separation segment and the unit inertia segment, the simplest convention to make is to take the unit inertia segment such that its conjugate is equal to the unit separation segment.

More generally we may take the unit inertia segment such that

$$\text{(conjugate of unit inertia segment)} = v\ \text{(unit separation segment)},$$

where v is a constant afterwards to be identified with what we call the "velocity of light."

Now the element N lies in W and is determined by three coordinates, say x_1, y_1, z_1, taken parallel to OX, OY, OZ respectively in the usual manner.

Further

$$\text{segment } NA = \text{segment } OM,$$

and so if t_1 be the length of OM in terms of the unit inertia segment, then the element A will be determined by the four coordinates x_1, y_1, z_1, t_1.

Let the length of the segment ON be denoted by a.

Then as in ordinary geometry

$$a^2 = x_1^2 + y_1^2 + z_1^2.$$

Thus if OA should be an optical line we must have

$$a^2 = v^2 t_1^2,$$

or

$$x_1^2 + y_1^2 + z_1^2 - v^2 t_1^2 = 0 \quad \text{..................(1)}.$$

Again, if OA should be a separation segment, and if r_1 be its length, it follows from the analogue of the theorem of Pythagoras for this case that

$$a^2 - v^2 t_1^2 = r_1^2,$$

or

$$x_1^2 + y_1^2 + z_1^2 - v^2 t_1^2 = r_1^2 \quad \text{..............(2)}.$$

Finally, if OA should be an inertia segment and $\bar{r}_1$ its length, it follows from the corresponding analogue of the Pythagoras theorem that

$$a^2 - v^2 t_1^2 = -v^2 \bar{r}_1^2,$$

or

$$x_1^2 + y_1^2 + z_1^2 - v^2 t_1^2 = -v^2 \bar{r}_1^2 \quad \text{...........(3)}.$$

Thus from (1), (2) *and* (3) *it follows that the expression*

$$x_1^2 + y_1^2 + z_1^2 - v^2 t_1^2$$

is positive, zero, or negative according as OA is a separation line, an optical line, or an inertia line.

If A be *after* O, it is clear from the convention which we have made that t_1 must be positive.

Thus the conditions that A should be *after* O are

$$\left.\begin{array}{ll} (1) & x_1^2 + y_1^2 + z_1^2 - v^2 t_1^2 \text{ is zero or negative} \\ \text{and} \quad (2) & t_1 \text{ is positive} \end{array}\right\}.$$

The conditions that A should be *before* O are similarly

$$\left.\begin{array}{ll} (1) & x_1^2 + y_1^2 + z_1^2 - v^2 t_1^2 \text{ is zero or negative} \\ \text{and} \quad (2) & t_1 \text{ is negative} \end{array}\right\}.$$

The conditions that A should be neither *before* nor *after* O are either that

$$A \text{ is identical with } O,$$

$$\left.\begin{array}{ll} \text{in which case} & x_1 = y_1 = z_1 = t_1 = 0 \\ \text{or else} & x_1^2 + y_1^2 + z_1^2 - v^2 t_1^2 \text{ is positive} \end{array}\right\}.$$

If (x_0, y_0, z_0, t_0) and (x_1, y_1, z_1, t_1) be the coordinates of elements which we shall call A_0 and A_1 respectively, we have simply to substitute $(x_1 - x_0)$, $(y_1 - y_0)$, $(z_1 - z_0)$, $(t_1 - t_0)$ for x_1, y_1, z_1, t_1 in the expressions obtained above in order to obtain the length of the segment A_0A_1 or to give the conditions that A_1 should be *after* A_0, or *before* A_0, or neither *before* nor *after* A_0.

It is to be noted that if

$$(x_1 - x_0)^2 + (y_1 - y_0)^2 + (z_1 - z_0)^2 - v^2 (t_1 - t_0)^2 = 0,$$

and the elements A_0 and A_1 are distinct, we are to take this as the condition that A_0 and A_1 lie in one optical line, and not, strictly speaking, that the segment A_0A_1 is of zero length.

We have already pointed out the peculiarity of optical segments, and how it is only possible to compare their lengths if they lie in the same or parallel optical lines, and we can now see how the analysis deals with this feature.

Now the condition that two distinct elements lie in an optical line gives us also the condition that the one should lie in the α sub-set of the other.

Thus if (x_0, y_0, z_0, t_0) be the coordinates of an element A_0, the equation of the combined α and β sub-sets of A_0 is

$$(x - x_0)^2 + (y - y_0)^2 + (z - z_0)^2 - v^2 (t - t_0)^2 = 0.$$

The α sub-set of A_0 will then consist of all elements for which this equation is satisfied and for which $t - t_0$ is zero or positive:

while the β sub-set of A_0 will consist of all elements for which the equation is satisfied and for which $t - t_0$ is zero or negative.

The set of all elements whose coordinates satisfy the above equation will be called the *standard cone* with respect to the element whose coordinates are (x_0, y_0, z_0, t_0).

Taking v equal to unity for the sake of simplicity, it is evident that the equation

$$x^2 + y^2 + z^2 - t^2 = c^2,$$

represents the set of elements such as A, where OA is a separation segment whose length is c.

Similarly the equation

$$x^2 + y^2 + z^2 - t^2 = - c^2$$

represents the set of elements such as A, where OA is an inertia segment whose length is c.

If we put $y = 0$ and $z = 0$ in the first of these we obtain

$$x^2 - t^2 = c^2$$

which gives us the relation between x and t for the portion of the corresponding set which lies in the acceleration plane containing the axes of x and t.

This then represents the analogue of a circle in the acceleration plane.

Similarly for the case of inertia segments putting $y = 0$ and $z = 0$, we get

$$x^2 - t^2 = - c^2.$$

The two equations

$$x^2 - t^2 = c^2 \quad \text{and} \quad x^2 - t^2 = - c^2$$

are of the same form as the equations of a hyperbola and its conjugate in ordinary plane geometry.

The equation

$$x^2 - t^2 = 0,$$

along with $y = 0$ and $z = 0$, represent the two optical lines through the origin in the same acceleration plane and these correspond to the common asymptotes of the hyperbolas.

It is now possible to express the various results which we have obtained in coordinate form and treat the subject from the analytical standpoint.

INTERPRETATION OF RESULTS

It is evident that any element whose coordinates are $(a, b, c, 0)$ must lie in the separation threefold W and accordingly the three equations

$$x = a, \quad y = b, \quad z = c$$

must represent an inertia line normal to W and therefore parallel to or identical with the axis of t.

Again, any equation of the first degree in x, y, z together with the equation $t = 0$ will represent a separation plane in W, while any two independent but consistent equations of the first degree in x, y, z, together with the equation $t = 0$, will represent a separation line in W.

Thus any equation of the first degree in x, y, z (leaving out the equation $t = 0$) will represent a rotation threefold containing inertia lines parallel to the axis of t; while any two independent but consistent equations of the first degree in x, y, z will represent an acceleration plane containing inertia lines parallel to the axis of t.

Thus corresponding to any theorem concerning the elements of W, there will be a theorem concerning inertia lines normal to W and passing through these elements.

Conversely, if we consider the system consisting of any selected inertia line together with all others parallel to it, then any two such inertia lines will determine an acceleration plane, while any three which do not lie in one acceleration plane will determine a rotation threefold.

Since these inertia lines must all intersect any separation threefold to which they are normal, it follows that they have a geometry similar to that of the separation threefold and therefore of the ordinary Euclidean type.

If then we call any element of the entire set an "instant"; any inertia line of the selected system a "point"; any acceleration plane of the selected system a "straight line"; and any rotation threefold of the selected system a "plane"; we can speak of succeeding instants at any given point and have thus obtained a representation of the space and time of our experience in so far as their geometrical relations are concerned.

The distance between two parallel inertia lines of the system will naturally be taken as the length of the segment intercepted by them in a separation line which intersects them both normally. This then will be the meaning to be attached to the *distance between two points*.

Time intervals in the usual sense will be measured by the lengths of segments of the corresponding inertia lines: that is to say, by differences of the t coordinates.

Since we have defined the equality of separation and inertia segments in terms of the relations of *after* and *before*, and have assigned an interpretation to these, it follows that, if this interpretation is correct, the equality of length and time intervals in the ordinary sense is rendered precise.

It is to be noted that the formal development of the theory of conical order does not in itself require that the α and β sub-sets should be determined by optical phenomena, but merely that there should exist some physical criterion of *before* and *after* such that the relations denoted by these words should satisfy our postulates.

Accordingly, if it should be found that some other influence than light possessed these properties, we should merely require to substitute this influence for light and interpret our results in terms of it.

The important point is that our theory gives a method of setting up a coordinate system on a purely descriptive basis and of introducing measurement without employing anything but the relations of *before* and *after*.

If we have got one coordinate system with a definite physical meaning, we can introduce any number of others.

The simplest is of course another system of the same kind as the first, and naturally it will be quite on a par with our original system of coordinates.

The distinction between different systems of this kind is, that while two parallel inertia lines represent the time paths of unaccelerated particles which are at rest relative to one another; two non-parallel inertia lines represent the time paths of unaccelerated particles which are in motion with uniform velocity with respect to one another.

Thus our geometry makes no absolute distinction between rest and uniform motion since any inertia line considered by itself is on a par with any other inertia line.

It is possible of course to introduce coordinates which are various continuous functions of x, y, z, t, but it is to be noted that these have a meaning only in virtue of the meaning which we have already assigned to our original system of coordinates in terms of *before* and *after*.

The change from one system of coordinates to another is equivalent to re-naming our set of elements, and may be compared to the translation from one language to another.

The four numbers x, y, z, t constitute a name for an element, and if we take four functions of these variables such that x, y, z, t may be expressed in terms of them, then these four functions may be regarded as constituting another name for the element, in a different language so to speak.

If we had a polyglot dictionary, it would be of little use to us unless we knew at least one of the languages which it contained, and a transformation from one system of coordinates to another without knowing the physical significance of at least one of the systems, leaves us in a similar situation to that which we might imagine an early Egyptologist would have been in who examined the Rosetta stone, but had no knowledge of Greek.

We have made use of the properties of light to give a physical interpretation to our postulates, but, if this be only approximate, it may be necessary to find some slightly different interpretation of *before* and *after*, or it may be necessary to somewhat modify our postulates, but in any case, some such analysis as we have outlined above appears to be necessary before we are at liberty to introduce numerical coordinates in this subject.

APPENDIX

If we assume some definite interpretation of the *before* and *after* relations, which is not necessarily identical with the optical one which we have provisionally supposed, we are able to introduce measurement and to show that, for an inertia line, we have

$$ds^2 = dt^2 - dx^2 - dy^2 - dz^2.$$

If now we change our system of coordinates and put

$$\begin{aligned} x &= r \sin\theta \cos\phi, \\ y &= r \sin\theta \sin\phi, \\ z &= r \cos\theta, \end{aligned}$$

then (r, θ, ϕ, t) constitutes a new name for the element formerly denoted by (x, y, z, t) and we get

$$ds^2 = -dr^2 - r^2 d\theta^2 - r^2 \sin^2 \theta \, d\phi^2 + dt^2.$$

Now let a modified measure of interval be introduced such that the measure $d\bar{s}$ of an indefinitely small interval is related to the corresponding unmodified measure by the equation

$$d\bar{s}^2 = ds^2 \left\{ 1 - \frac{2m}{r - 2m} \left(\frac{dr}{ds} \right)^2 - \frac{2m}{r} \left(\frac{dt}{ds} \right)^2 \right\}.$$

Then
$$d\bar{s}^2 + \frac{2m}{r - 2m} dr^2 + \frac{2m}{r} dt^2 = ds^2.$$

Thus

$$d\bar{s}^2 = -\frac{1}{\left(1 - \frac{2m}{r}\right)} dr^2 - r^2 d\theta^2 - r^2 \sin^2 \theta d\phi^2 + \left(1 - \frac{2m}{r}\right) dt^2.$$

This is the form given by Schwarzschild for the region round a single spherical body on Einstein's gravitation theory, and a similar method may be employed in other cases.

The problem is very analogous to that of the brachistochrone, in which the modified measure of interval is the time taken to traverse it.

The condition that the modified measure of an interval whose ends are at a finite distance apart should be stationary determines a path from the one to the other, which, in general, will not be a straight line in the original coordinates.

In the general case suppose that we have any four coordinates x_1, x_2, x_3, x_4, which have a definite interpretation in the simple theory as built up from the *before* and *after* relations, and suppose that for an indefinitely small interval we have

$$\begin{aligned} ds^2 = {} & h_{11} dx_1^2 + h_{22} dx_2^2 + h_{33} dx_3^2 + h_{44} dx_4^2 \\ & + 2h_{12} dx_1 dx_2 + 2h_{13} dx_1 dx_3 + 2h_{14} dx_1 dx_4 \\ & + 2h_{23} dx_2 dx_3 + 2h_{24} dx_2 dx_4 + 2h_{34} dx_3 dx_4, \end{aligned}$$

and suppose we wish to derive a new system such that

$$\begin{aligned} d\bar{s}^2 = {} & g_{11} dx_1^2 + g_{22} dx_2^2 + g_{33} dx_3^2 + g_{44} dx_4^2 \\ & + 2g_{12} dx_1 dx_2 + 2g_{13} dx_1 dx_3 + 2g_{14} dx_1 dx_4 \\ & + 2g_{23} dx_2 dx_3 + 2g_{24} dx_2 dx_4 + 2g_{34} dx_3 dx_4. \end{aligned}$$

In order to do so we have only to assume that $d\bar{s}$ is related to ds in such a way that

$$d\bar{s}^2 = ds^2 \left\{ 1 + (g_{11} - h_{11}) \left(\frac{dx_1}{ds}\right)^2 + (g_{22} - h_{22}) \left(\frac{dx_2}{ds}\right)^2 + (g_{33} - h_{33}) \left(\frac{dx_3}{ds}\right)^2 + (g_{44} - h_{44}) \left(\frac{dx_4}{ds}\right)^2 + 2\,(g_{12} - h_{12}) \frac{dx_1}{ds}\frac{dx_2}{ds} + \ldots\ldots \right\}.$$

It thus appears that these complicated systems of geometry may be constructed from *before* and *after* relations, provided we make use of a modified measure of interval.

If it be convenient for some purpose to describe the paths of particles as geodesics in some of these complicated geometries, there is no particular reason why we should not do so; but this does not imply any "curvature of space."

I put forward the following suggestion as the book is going to press, though I cannot claim to have worked out its full implications:—

If, with Einstein, one takes a particular quadratic form as corresponding to a particular gravitation field, we might suppose that a change in the gravitation field produced by something else than gravitation (say, for instance, a change which was initiated by an act of will on the part of a living creature) corresponds to a change in the quadratic form.

If it be permissible to take such change as the influence which gives the *before* and *after* relations their physical significance, instead of using optical phenomena for this purpose, then we might suppose that the *before* and *after* relations, as formulated, hold in some sense whether matter be present or not, but that it is only in the absence of appreciable quantities of matter that the postulates can be interpreted strictly optically.

Thus the original four-dimensional manifold would appear as the essential thing, while these complicated geometries would be regarded as analytical developments useful for special purposes.

Lightning Source UK Ltd.
Milton Keynes UK
UKHW012042140220
358775UK00001B/1